Engineering Philosophy

Engineering Philosophy

Louis L. Bucciarelli

DUP Satellite / 2003

DUP Satellite is an imprint of

Delft University Press
P.O. Box 98
2600 MG Delft
The Netherlands
Telephone:+31 15 2785678
Telefax: +31 15 2785706
E-mail: Info@Library.TUDelft.NL

ISBN 90-407-2318-4

Printed in the Netherlands

Table of Contents

1

Introduction

"Let's stop all this philosophizing and get back to business"[1]

Philosophy and engineering seem worlds apart. From their remarks, we might infer that engineers value little the problems philosophers address and the analyses they pursue. Ontological questions about the nature of existence and the categorial structure of reality – what one takes as real in the world – seem to be of scant interest. It would appear that engineers don't need philosophy; they know the difference between the concrete and the abstract, the particular and the universal – they work within both of these domains every day, building and theorizing, testing and modeling in the design and development of new products and systems. Possible worlds are not fictions but the business they are about. As Theodore Von Karman, an aerospace engineer and educator, reportedly claimed

> Scientists discover the world that exists; engineers create the world that never was.

Epistemological questions about the source and status of engineering knowledge likewise rarely draw their attention.[2] Engineers are pragmatic. If their productions function in accord with their designs, they consider their knowledge justified and true. Such knowledge, they will show you, is firmly rooted in the scientific explanation of phenomenon which, while dated according to physicists, may still provide fertile grounds for innovative extension of their understanding of how things work or might work better. This complicates my task; for my intention is to show that philosophy can matter, does matter, to engineers. I want to explore in what ways it might contribute to doing a better job of designing and, as a teacher of the craft how it might help us in better preparing our students for life as well as for professional practice.

Most professionals would agree that the process of designing itself stands in need of improvement.[3] Many recognize that the days when an engineer could work alone in his cubical on some specialized bit of the whole, then throw his work over the wall to the next in line, are over. Some, in attempting to improve engineering education, are even challenging the traditional lecture format, where faculty, back to their students, cover the chalk board, unfurling the fundamentals of their specialty with little regard to the worlds around them. Yet the remedies and changes proposed to renew the profession on both counts fall short. I hope to explain why this is the case.

I also hope that this exploration, by getting behind the Oz-like curtain draped out front, will give the general reader a better and truer understanding of engineering – what resources, knowledge and know-how it requires, what it can not, as well as can do. My remarks might also prompt some interest on the part of philosophers; perhaps they will see in engineering thought and practice a fertile field for the cultivation of old themes but in a new light.

Certain philosophers *have* considered technology a worthy subject for comment and analysis. Few, however, have considered the design of technology a subject worth addressing, perhaps because of the inaccessibility and complexity of the design process or because engineering is deemed so mundane and rule-respecting that there is nothing worth critique and analysis – a presumption which resonates with the character of much philosophical exploration when technology is the target, e.g., explorations of the autonomous character of technique and its "impacts".

At the Technical University of Delft, there is a group of philosophers who see technology otherwise: Funded by the Dutch equivalent of the US National Science Foundation, they seek to better understand the "dual nature" of technological artifacts.

> On the one hand, these are *physical objects* with a specific physical structure (physical properties), the behavior of which is governed by the laws of nature. On the other, an essential aspect of any technological object is its *function*....
>
> This dual nature of technological objects is reflected in two different modes of description, via. a *structural* and a *functional mode*. Insofar as it is a physical object, a technological object can be described in terms of its physical (structural) properties and behavior....free of any reference to the function of the object....With regard to its function, a technological object is described in an intentional (teleological) way:[4]

While the distinction is real, the duality ought not to be taken as expressing a disjunction of the sort reflected in those who talk of "two cultures" or of "impacts of technology on society". The issue at hand is not of this sort. Rather the quest is to explain how participants in product design and development transform interest, beliefs and intentions into a functioning product. How do the authors of an artifact or system endow (material) substance with the appropriate form so that it will function properly, as intended? This, too, is my concern.

I am not a philosopher: All of my formal education has been in engineering. I have, however, over the past thirty five years, taken seriously the challenge set by C.P. Snow and worked to bridge the two cultures. I know something about the history of science and have taken a keen interest in the social study of science and technology. Still, philosophy stands a world apart, even from these domains. It is another scholarly discipline, a whole other world requiring new learning, a new vocabulary, a new sense of what is a legitimate question – none the less what is an important question – and what constitutes a coherent, legitimate response. I am just beginning to learn to speak in this domain. And like a neophyte, I have many awkward questions. My approach, then, is not to try to construct a formal philosophical treatise, but to draw upon my experiences – as designer, as consultant, as researcher, as teacher – in setting out and exploring some hopefully fruitful connections.

Most of my professional life has been spent as a faculty member doing research and teaching within the field of engineering mechanics. One thing that continues to surprise me each year, is the apparent inability of some of my students, and there are always some, to see the world – of bridges and buildings, forces and torque – as I do. At times, they come up with the most bizarre questions when challenged with certain problems and yet seem unable to accept and digest the explanations I provide[5]. They have, what faculty of schools of education responsible for the preparation of science teachers tag "serious misconceptions" – a topic about which there is an ever-growing body of literature in the scholarly journals in science education. The "naive science", or "common sense science", or, still more tolerant, "alternate world views" of youth are all to be rooted out, washed away, to make room for the way things really work, e.g., the true stories about planetary motion; equal and opposite internal forces; uniformly accelerated motion, and the like.

I myself, find this deviant behavior on the part of my students refreshing and provocative: When so challenged, I want to know where in the world they got this strange way of seeing things. What do they call upon to justify their misconceptions? If I can reconstruct something of the student's conceptual scheming – which he or she might or might not acknowledge as their way of thinking – then I have a much better chance of success at conversion of the student from error to my way of seeing.[6]

Misconceptions are not necessarily disabling. Common sense ordinarily serves us well as a basis for thinking, acting and social exchange (until its undoing by science). Unquestioned presumptions on the one hand and long dead myth and metaphor on the other, are normally harmless. In fact, in ordinary times, they are enabling.[7] Indeed, if misconceptions and common sense were somehow disallowed, we would still be living in a stone age. Popper is right: Progress is the product of ill-conceived conjecture and its possible refutation.

But these are not ordinary times. Advances in technology, particularly in computer, communication and information processing technologies have swept over and shaken up the world of politics, commerce, business, engineering and even engineering education. Engineers are both responsible, in part, for the development of this technology and are subject to it, must learn to live with it and to put it to use effectively as much as any other persons. Faculty must learn to teach about and with it. Practitioners must both cope with it and shape it to their immediate needs.

The changes afoot go beyond learning to put these new "tools" to use on the job for there is an across the board upheaval in the nature and organization of engineering practice and of professional life. Computation tools and methods have become ever more sophisticated and powerful so the emphasis at work shifts from calculation and analysis to model making, to world making. Lifting our heads up from the calculator and the drawing board, our field of view enlarges; what were distant shadows now become features to be reckoned with. Design criteria broaden. Industrial ecology means that boundaries around the product can no longer be so impervious to the interests of "outsiders". "Open software" design explicitly recognizes the legitimacy of others, down stream to contribute to the design, calling into question the idea of a "finished product" as well as challenging traditional norms governing "ownership". Ethical and safety questions permeate the design process, seeping through into the foundational structure of scientific and instrumental analysis. At the same time, engineers are to be multi-disciplinary, polyvalent; they are to be able to work in teams – and not just teams meeting in the same room. For, in this day and age, participants in a product's design and development may be distributed around the globe. Design is to be done concurrently with many voices in harmony. I am not alone in emphasizing the uncertainty and ambiguity engineers must cope with as a result of this new mix.

It is at times like these, when our ordinary ways of thinking and doing are called into question, when philosophy might prove enlightening. For philosophy's aim is to clarify, to analyze, to probe and explore alternate ways of seeing, of speaking, and, ultimately, of remaking the world. These essays are meant as a go at evaluating certain characteristics of engineering thought and practice with the aim of bringing to the surface their essentials – the essential, fundamental beliefs of what may be called an engineering mind set. The hope is that this will help us see our way clear through these times of change, to rationally address what needs changing, what's best left sacred.

I tried once before to sketch out the fundamentals. In *Designing Engineers*[8] I noted that the worlds engineers fabricate and work within show hierarchy. Mathematical/scientific theory is higher up; e.g., the concepts and principles of the theory of elasticity are more fundamental than the specialized relationships which derive from the theory and describe the behavior of particular phenomena – the stresses within an end-loaded, cantilever beam, the buckling load of a column, the modes of vibration of a thin, flat plate. I spoke of an engineer's commitment to continuity, to conservation principles, to cause then effect, to both concrete particulars and abstractions; to measurement and quantification, estimation and certainty. Here I take another stab at this, singling out for reflection certain characteristics of what I there dared to label an engineer's "cosmology".

Foremost among these is the reductive way of seeing the world, of framing the task. An instrumental, usually quantitative, assessment of the functioning of a design or system is, of course, necessary to getting the job done, but it does not suffice. Consider, for example, the reduction of a design task to clearly demarcated, independent subtasks. While the attempt should always be made, I claim that complete independence of one task from another can not be achieved; negotiation at the "interfaces" of the different subtasks will always be required. If the possibility, nay probability, of this sort of social exchange goes unacknowledged or does not even enter one's field of view, any attempts at improving the design process are bound to be discouraging. For it is a mistake to imagine that the tensions springing from the different proposals of participants in design which precipitate out at the interfaces can be resolved by instrumental means, e.g., an optimization algorithm, alone.

Or consider the assumption "all other things being equal" – always made at some point in an engineer's seeing and thinking, modeling and predicting. Of course, one ultimately has to attribute a measure of uniformity to the world and limit one's attention to what is taken to be of primary importance to the successful functioning of one's plans and productions but where boundaries are drawn, what's in, what's out, depends as much upon the categories one acknowledges to exist as it does upon the relative significance of factors accommodated so "naturally" from the instrumental perspective of one's speciality. For example, if you see social agents as machines, as ergonomic objects or behaviorist's boxes, then it matters little how much further articulation of the machinery is pursued if someone asks how the product will contribute to a sense of community among the citizenry. Here is misconception of a different kind, akin to blindness.

Ceteris paribus and reduction are fundamental to engineering thought and practice. Allied is a more prosaic notion, that of "product". We envision the boundaries of a product as sharp and distinct. The artifact is not only detached in a material sense but taken as neutral and value free. This too is myopic. A product is in some measure material, made to fulfill some intended function, but is better construed as, in another sense, ideological as well – a human creation which reaches out beyond the box it came in to enable and to affect (and infect) our thoughts, our values, our beliefs as well as our practices. We all might agree that technology has the dual nature of structure and function, but even this way of speaking is too limiting: It presumes that "structure" contains the hard science, there for all to see in the same way, while function is limited to the acknowledged intentions of designers - and possibly users.

My aim is to subject these characteristic, fundamental ways of seeing, or not seeing as the case may be, to critique. In what way are they no longer justified? What new forms of thinking, new ways of seeing, are required?

As engineering faculty, we claim to teach the fundamentals. Used in this more ordinary way, the word points to the primary concepts and principles that lie at the roots of the disciplines, whether mechanics, electronics, thermodynamics, etc. (Different disciplines, different species of trees, some with very shallow roots, others which go all the way to China – from a Western perspective). These underwrite the engineer's heuristics and derivation of relationships among variables and parameters which describe, in what I call object world language, the way a particular product, artifact, system will hopefully function. So scattered about in the chapters which follow are equations and simple abstract images. Confronting these, I encourage the reader not to turn away but to read on, for this essay is not meant as an engineering textbook. Rather, I hold that to analyze and explain the status and function of engineering knowledge requires a display and critique of the texts engineers themselves construct and rely upon as much as it does explaining the rationale of their productions.

My concern with texts reflects a shift in scholarship over the past century within the humanities, arts and social sciences. Philosophy, it is said, has taken a "linguistic turn". This appears to mean that attention of scholars has shifted from an analysis of the facts of the world (including persons) found in the world "out there" or "inside me" to an analysis of the way we speak about the world. In so doing, philosophers have discovered that much of the time we don't know what we are talking about and have strived to clarify, to show us how what we take as problematic in the world is, or may be, but a problem with the way we express matters. More recently, some claim that philosophy has taken a "cognitive turn" or at least a branching. This appears to mean that the study of language is not enough; we are to go inside now and analyze the cognitive structures we construct, project, and express in language as we muddle along.

From the tone, you might surmise that I do not fully agree with all of this manuevering. True, I don't fully agree but I do embrace the notion that to explain the world we have to pay close attention to who, in this case engineers, is doing the explaining – their predilections and presumptions, the traditions they draw upon, the way they describe the world they live in. It is not enough to study propositions, facts, abstract and in themselves alone; context matters. Wittgenstein, after turning away from a rather sterile picture of the world of facts and propositions, recognized that rules and meanings in talk and texts were mixed up with practice, with context of use. In order to understand how a rule is to be applied, he claimed, one must study how it is used. Those who study "situated cognition" appear to be saying much the same thing, i.e., the meaning of statements is embedded in practice[9].

Some critics see danger in all this movement away from classical philosophical concerns. If language, if practice matters and the import of our rules, the games we engage, the things we say, and the texts we write can only be explained by paying full attention to context, including the intentions of the author and the culture of the reader, then the established boundaries among scholarly disciplines tend to dissolve, or at least be challenged. A proposition in the abstract, like an isolated fact, looses its significance. The narrative within which these are embedded need to be studied. Rhetoric and philosophy are joined.

There is a resonance here with the fact/theory dichotomy in science. Philosophers and historians of science have come to recognize that facts are conditioned by theory and theory does not stand apart from prevailing, more public and general ways of seeing and talking about the world. Facts are not just out there in the world, waiting to be discovered or uncovered. Theory, extant beliefs, the stories we tell, as well as our instruments for seeing, fix what we see as something important – what are significant things, variables, and parameters – as well as how they interact and relate in the world. Thomas Kuhn went so far to claim that a Copernican lived in a different world than his Aristotelean predecessor.

Engineers do construct narratives, not very fancy ones perhaps, but subtle in ways I hope to explain. It is quite common to think that the explanations of engineers are objective, scientific, one-dimensional, lacking in ambiguity, trope or metaphor. To see that it is otherwise, requires we pay full attention to the way instrumental reasoning is embedded in text and to allow that the form of argument and the language used, in short, the rhetoric of engineering is part and parcel of instrumental explanation. Note: In this I am speaking of the ordinary texts engineers write and speak in their day to day work. I am not going to say much about the language of their grant applications or stories manufactured for public consumption.

Another shortsightedness, namely that the past is irrelevant to doing the work of today, derives from the apparent timeless quality of engineering explanation. Engineering explanation is like scientific explanation in this respect. A machine, a new product, a computer program – whatever artifact one has in view – pretends to work always and everywhere in the same way. The principles upon which it relies are expressed as they were in the beginning, are now, and ever shall be, forever and ever. The constitution of machinery needs no new interpretation with changing times. So history, like rhetoric, is irrelevant to understanding technology – or so it would seem.

One can indeed construct an explanation of engineering thought and practice from this perspective, limiting your field of view to include only those events and "variables" that have this timeless, instrumental quality. One can, for example, study the way a new products works, lay out the reasons why it works and how the integration of its different parts and subsystems is achieved, and claim that this instrumental explanation describes as well the design process, the way it came to be. This is a mistake. It fails to acknowledge that designing is a social process of negotiation, of iteration, of rectifying mis-steps, even misconceptions – a process rich in ambiguity and uncertainty.

History is important to understanding the genesis of technology. In attending to history we get behind, as it were, technology's so dominate instrumental presence to explore the give and take of ideas, the mix of material representation and scientific principle, and the field of constraints within which the engineer labors. We uncover both the dated nature of technique and reveal the conditions for its making.

My attempts to explore and explain the connections and relationship between philosophy and engineering, through the study of engineering narrative and historical process, is a way of acknowledging that context matters and must be attended to if we are to say anything that takes us beyond an instrumental explanation of the type engineers profess themselves. In these times of change, if we are to claim some control over the future, we must allow that what needs to change includes more than the tools, the organization, the methods, the hardware and software, but more fundamentally, ways of perceiving and reading the world.

In engineering we see the world through glasses that let through the instrumental, the calculable, the scientific, alone; the rest of the world is but a haze. My claim is that we *do* need to see the world differently. The world we live in, are remaking now, is a different world. The reductive, instrumental character of engineering thought and practice is what we seek to explain and critique.

In the essays that follow, we approach our topic, not directly, but through different kinds of activities engineers engage: The first essay, Chapter 2, describes the languages of design and the negotiations their differences entail. The second addresses how engineers deal with failure and error. The third explores the ways they model and idealize the world they remake. The fourth how they teach.

Notes.

1. Remark recorded at a meeting held to evaluate possible design options.

2. A notable exception is the work of Walter Vincenti, another aerospace engineer! Vincenti, W., *What Engineers Know and How They Know It,* Johns Hopkins Univ. Press, 1990. Less well known is an excellent article by Vincent Hendricks, Arne Jakobsen, and Stig Pedersen, "Identification of Matrices - in science and engineering", *Journal of General Philosophy of Science*, 2, 2000.

3. It is an interesting question whether improving the process of design will necessarily improve the quality of the product. This need not be the case, though one would expect it might be so. The question, however, is hazardous, and this in two ways: First, what is judged a better process or product will depend upon whom you ask. Second, the whole argument can easily become circular, especially if judgement is passed after the fact, after the better or worse product goes out the door and is launched out into the world.

4. *The Empirical Turn in the Philosophy of Technology,* Kroes, P. & Mejers, A., (eds). Elsiver Science, 2000. p. 28.

5. I conjecture that it was this sort of experience that Wittgenstein had while teaching middle school students off in the hinterlands of Austria that provoked his philosophical transformation. Monk, Ray, *Ludwig Wittgenstein: the duty of genius* London, Jonathan Cape, 1990.

6. Of interest too is a more general question: How does one, how can one construct a rational explanation for what is apparently irrational explanation?

7. Lakoff, G., and Johnson, M., *Metaphors We Live By,* Chicago, Univ. Chicago Press, 1980.

8. Bucciarelli, L.L., *Designing Engineers,* MIT Press, 1994.

9. I agree, but I am put off a bit by the emphasis upon "practice". It smacks too much of un-reflective action, of the exercise of skill alone without thinking. Certainly a skilled practitioner in action – a carpenter, plumber, surgeon, or pianist, for example – is ordinarily not doing much thinking; the exception being when the rhythm of execution is challenged by the unanticipated. That's the whole point of being skilled; you don't stop after each impact of the hammer, each pass of the torch or knife, to reflect upon what you have just accomplished or where the next note on the keyboard lies. Engineers are skilled too in different ways; but they *do* stop and think, and rethink, and redo, and rethink as they go about designing.

2

Designing, like language, is a social process.

There is a cartoon familiar to most aerospace engineers which purports to depict the design of an aeroplane: It shows in some half dozen frames on a single page, the different visions of the final product which accord with the different interests of those responsible for its design. The vision of the structural engineer includes massive I-beams which assure the craft does not fall apart; the vision of the design participant responsible for powering the craft shows very little structure other than that required to support the huge twin engines. The aerodynamicist's representation is as sleek and slim as one might imagine; there is hardly room for the pilot. And so it goes; and, indeed, that is akin to the way it goes. For in a nut-shell, engineering design is a process which engages different individuals, each with different ways of seeing the object of design but yet individuals who in collaboration, one with another, must work together to create, imagine, conjecture, propose, deduce, analyze, test and develop a new product in accord with certain requirements and goals.

Participants in any design project of all but the simplest kind, working in different domains on different features of the system, will have different responsibilities and more often than not, the creations, findings, claims and proposals of one individual will be at variance with those of another. While they all share a common goal at some level, at another level their interests will conflict. As a result, negotiation and "trade-offs" are required to bring their efforts into coherence. This, in turn, makes designing a social process. If we stop here, this not terribly problematic. What complexifies the situation and makes designing a challenge of the highest order is that each participant sees the object of design differently.

In *Designing Engineers*,[1] I provide evidence for this claim: I report on three design projects, one which consisted of a small group, on the order of ten people, engaged at a firm in the design and development of a large photovoltaic module. In the book I describe how participants in design saw the design differently, then analyze the consequences of my observations and conjecture.

The team at "Solaray" included a mechanical engineer responsible for the design of the module frame, the protective layers of backing, the sizing of the cover glass, and assembly of the product. An electrical engineer was ultimately responsible for the design of the series/parallel circuitry of the photovoltaic cells, of the number and placement of diodes, and of the choice of electrical hardware including junctions and cabling.

The responsibility of a materials person from cell production was to characterize the performance of the individual cells and explore how statistical variations from cell to cell would affect the output of the module as a whole. A marketing person with an engineering background headed up the system's engineering group. His primary concern was to see that the prescribed current-voltage characteristics of a single module would allow for the build-up of systems of aggregates of modules which could meet the special needs of a wide variety of different customers. Each of these people had some significant say about the design of the new, large photovoltaic module.

All, as members of the firm Solaray – a corporate entity whose purpose like others was production for profit – came together to design a product of quality which would contribute to their and the firm's survival. At the same time, they were in competition one with another, each evidencing a different view of the object of design and of the relative importance of the critical design parameters and of how they should be set.

To the systems engineer, the module was a "black box"; he viewed it as a functioning whole, a unit joined with others in series and parallel to provide the bus voltage and power level a range of systems might require. The materials person, concerned with the characterization of the distribution of properties of the photovoltaic cells didn't "see" a module at all; rather, to her, the cells were units in an electrical circuit whose overall behavior depended upon the degree of "mismatch" among the current-voltage characteristics of the individual cells.[2] The electrical engineer from the system's group on the other hand took the cells as identical in his design of alternative possible series/parallel circuit configurations. He described the module as a circuit topology of ideal current generators and associated electrical elements, e.g., diodes, connectors. The mechanical engineer focused on the module's structure; the frame and the cover-glass were foremost in mind. To him, the cells were fragile wafers of glass which needed to be supported, fixed in a plane and protected from the weather.

The differences in readings of the photovoltaic module correlate with the different individual's responsibilities and are rooted in their different educational histories and experiences on the job. Each individual projects out into the design process her or his own reading of the object. It's like a theory of vision of antiquity that shows rays emanating from the eye out onto the world, rays which reflect and return to the seer signaling the presence and nature of the object. Analogously, different individuals in design emit rays of a different character which are reflected back differently by the object of design. Each participant sees differently in accord with the standards of thought and practice within their domain of specialization. It's like they live in different worlds.

We speak occasionally of multiple worlds, e.g., "the world of mathematics" or "the world of algae", "the world of Escher", or "The Wonderful World of Insects".[3] The world of algae is different from the world of Escher, and insects have a whole different milieu to contend with. All these creatures live in the same world but we must also allow that they see and experience the world, the big world at large, out there, differently. We can even claim they live in incommensurable worlds. So too within engineering, the structural engineer has a certain way of dealing with, describing, and speaking about the world. He or she draws upon an infrastructure of standards and regulations, of suppliers and consultants for support, e.g., a library of standard structural forms and sections. More formally, there is a mathematical theory which describes at a fundamental level how elastic continua and structural elements behave when weighted down or flown, vibrated, or deformed. There are computer programs designed specifically for modeling complex and simple structures of all shapes and sizes. Certain machinery exists for testing materials and completed structures; instrumentation and sensors have been developed specifically for his or her use in this regard. structural engineers have their own professional journals, professional societies, and their own existential pleasures - I suppose.

The world of the electronics engineer is different. Different infrastructure of standards and regulations, different off-the-shelf devices to choose from and build with, different forms of mathematics, different ways of sketching and modeling – the block diagram figures largely here – and different computer tools for modeling their systems. Testing instruments and apparatus are different. Time even has a different quality. Dynamic response predominates. And of course their professional journals and societies differ from those of other engineering disciplines.

And so it is, down through the list of all those who have a significant say in a design process. Each inhabits a world of things particular and employs specialized modes of representation. A world with its own unique instruments, reference texts, prototypical bits of hardware, special tools, suppliers' catalogues, codes, regulations and unwritten rules. There are exemplars, standard models of the way things work from the disciplinary perspective of the particular world and particular metaphors which enlighten and enliven the efforts of inhabitants. There are specialized computational methods, specialized ways of graphically representing states and processes. And each participant works with a particular system of units and with variables of particular dimensions, certain ranges of values perhaps. Dynamic processes, if that is their concern, unfold with respect to a particular time scale - for someone's world it may be milliseconds, in another's, hours or days. I say that different participants work within different *object worlds*.

Engineering Design - Other Perspectives.

Before going further, I want to spend a moment and describe some other views, some more standard representations, of the engineering design process. I do this for several reasons: To describe design as a social process does not seem, at first, to take us very far if our intent is to come up with prescriptions for improving the process. Second, these other ways of viewing design process are taken seriously by those who seek to improve the process and are even put to use in practice. I ought to acknowledge that fact. Third, and of primary interest, keeping in mind our desire to connect up philosophy with engineering, comparing the view of design as a social process with the view of design as an instrumental process - the meaning of which will become clear with my examples - prompts questions about the ways of thinking and doing in engineering which, if we are to address them seriously, requires a philosophical sensitivity. My purpose, then, is not so much to compare these methods in terms of their efficacy for guiding design practice but rather to explore how they represent different ways of thinking about the design process.

Figure 2.1, from Dixon[4] shows a traditional picture of the design process as it is presented and described in the engineering textbook. It shows, on the left, some stages in the design process beginning with "Goal recognition", "Task specification", and ending with "Distribution, Sales, and servicing". The box labeled "Engineering Analysis" is expanded at the right to show its internal workings.

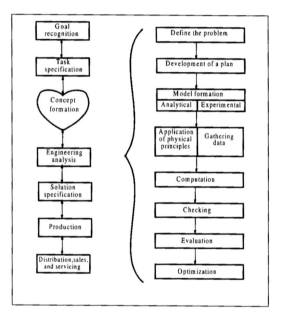

Figure 2.1 The Design Process

Figure 2.2, is another, similar in kind. The authors, Pahl and Beitz[5], note that

special emphasis is on the iterative nature of the approach and the sequence of the steps must not be considered rigid. Some steps might be omitted, and others repeated frequently. Such flexibility is in accordance with practical design experience and is very important for the application of all design methods.

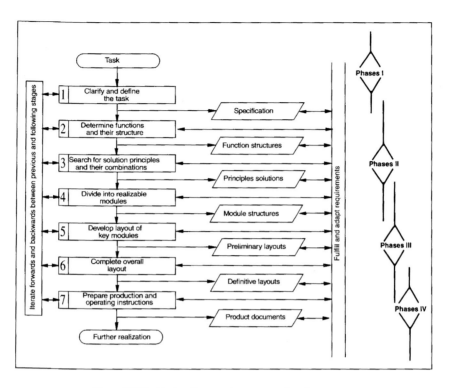

Figure 2.2 The Design Process: A more complex picture.

Finally, Figure 2.3, taken from a more recent mechanical engineering journal[6] shows a different sort of vision of design which in some respects is in accord with my claim that different participants, with different technical responsibilities and interests, see the object of design differently. The figure depicts "...a three discipline coupled system analysis". It shows what I would consider three object worlds; within each of the boxes labeled "disciplines" or "subspaces" I see persons responsible for the different "state variables" y_1, y_2, and y_3.

The authors note that "... these state variables are three independent sets" while the x's stand for variables which are to be chosen to ultimately optimize the design. Some of these, x_{sh}, are shared among the three disciplines. The g's are equations of constraint. The f's "...contain the design objectives of disciplines..." They depend upon the state variables.

These three representations of design process display similar characteristics: All but the last suggest that designing is a dynamic process, done in discrete phases; the similarity of Figures 2.1 and 2.2 to the block diagrams of the controls engineer is to be noted. But here, while there is feedback referred to as iteration, in contrast to the diagrams used in a controls analysis there is a definite start and end to the time-varying process. In Figures 2.1 and 2.2, time begins at the top of the diagrams and flows on as we move down through the process. There is the possibility of backstepping and as Paul and Beitz noted, we can even leave out some steps as we move along.

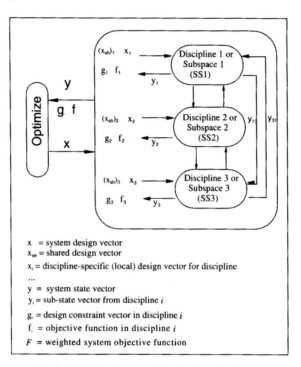

x   = system design vector
x_{sh} = shared design vector
x_i = discipline-specific (local) design vector for discipline
...
y   = system state vector
y_i = sub-state vector from discipline i
g_i = design constraint vector in discipline i
f_i = objective function in discipline i
F = weighted system objective function

Figure 2.3 Multi-objective Optimization in Design

The last diagram differs in important ways; the payoff is positioned at the left, not at the bottom; the flow of time is not acknowledged. And with good reason for this diagram does not really represent a design process at all but rather a purely instrumental process – an algorithm – for the determination of design variables and parameters in an optimum way. All of this is done in an instant by the machine. This is not to say that this is the actual way the process goes, i.e., as a one shot run of an optimizing algorithm. The advantage of machinery is that it can be run over and over again without wear and tear. Put to use in this way, its input, output and working assumptions can become the ingredients of exchange among the different parties to the design - assuming they exist.

We have here different ways of seeing and representing the engineering design process. Mine expressly champions its social nature - without a block diagram so far; a second embodied in the models of Dixon, and Pahl and Beitz emphasizing process but without people - an instrumental or mechanistic process; and a third which presents designing as a computational algorithm.

It might be argued that these three are meant to address three different dimensions of design activity: Mine concerned with the organization of the people charged with the design task, the instrumentalists concerned with the organization of the work of the design task, and the analysts with decision-making in the later stages of the process. There is something to this but there is more to the story. This way of slotting and distinguishing the three representations is in itself too instrumental. It smooths over the disjunction I seek to put in relief – that between design as a purely technical, instrumental process and design as a social process. To label my vision as organizational is short-sighted and misses the point.

In a way, this faulty characterization of what I am about *makes* my point. For I am very much concerned with the form of expression of our authors and participants in design. My attention to "the voice" of texts and diagrams is purposeful: A major concern, in this and subsequent essays, is the rhetoric of engineers at work in designing. My claim is that different forms of expression go hand-in-hand with different ways of thinking about the world, about the existence of conceptual entities – their ontological status – and about the meaning and scope of the principles and requirements of the different paradigmatic sciences that frame thought and practice within object worlds. My framing of design as a social process in which different participants work within different object worlds which, in some restricted sense are incommensurable worlds, leads me to claim they speak different languages.

Language(s) of Object Worlds[7]

Of course, participants in design share a common language, their native language, e.g., English. And this characterizes the sounds one hears and the words one reads standing aside any participant working within an object world. But the sense and meaning of these expressions heard, voiced and read ought not to be construed in ordinary terms. For object worlds are worlds where specific scientific/instrumental paradigms fix meaning. There our ordinary language is used in such a specialized way it is as if a participant is speaking a different language. Not different in the sense that for you, as a foreigner, a translation would make meanings clear; after all, the words *are* English, although a technical dictionary would not be without merit, but different in that knowing and understanding the concepts and ideas and relationships among the things of an object world require new learning, like the learning of a foreign language. And the challenge is not just a matter of coming to grips with, for example, the widespread use of mathematical symbolic expression but, just as getting it right in French or Dutch, Spanish or Japanese means being able to "live" the culture to a degree, so too within object worlds language is a matter of convention and custom, often curious practices and forms of expression as well as tokens and grammar, jargon and idiom.

Each object world language of an engineer is rooted in a particular scientific paradigm which serves as a basis for conjecture, analysis, testing and designing within that world. I have already referred to the world of the structural engineer who speaks of stress and strain; of displacement, stiffness, and load path. These terms have specialized meaning: stress is both a physical thing, as force per unit area, and a mathematical thing, as second order, symmetric tensor. It is like the common word stress, e.g., how you feel when under pressure, in that too much can lead to failure but your common knowledge would not enable you to size the beams of a bridge, an automobile chassis, or the wing of aeroplane. Stress, strain and displacement are variables; they are quantifiable; they bear specific units and dimensions, enter into relations derived from the parent scientific theory using appropriate kinds of mathematics.

I say that an object world language is a *proper language*: Specialized, yes, dialect, if you like, but more to my point - proper in the sense of the French word "propre": A language neat, clean, exact, all in order, honest - at least in its unadulterated state. It masquerades as the participant's natural language but understanding and meaningful expression within object worlds demands more of the native speaker than proficiency in his or her natural language. A proper language is technical, instrumental; it has the form of a scientific language, perhaps ornamented with more things of the world than a true scientist would welcome. It is learned on the job as well as in disciplinary course-work within schools of engineering. Its canonical form is codified in handbooks, standards and textbooks.

A proper language is in good measure analytic. Within the world of the structural engineer, once one defines stress and strain - the latter, like stress, a second order, symmetric tensor but physically the measure of deformation at points within a continuum - then insists upon equilibrium of internal and external forces in accord with Newton Laws, upon continuity of displacement, and finally inserts some parameters to stand in for the elasticity of the material one can, in principle, derive precise statements that fix the stress levels in the beams of the bridge, the specification-meeting shape of the members of the automobile chassis, and the frequency of vibration of the fluttering wing of an airplane. I say "in good measure" because in application, an individual's experience and traditional ways of doing things enter at many stages shaping and fixing presumptions, assumptions, heuristics and approximations - and this in the crafting of prototypes and setting of test methods as well as in formulating a crude or sophisticated mathematical model. As with Gilbert Ryle,[8] *knowing how* is just as important as *knowing that* within object worlds. Still the prevailing mind set is analytic, a fact displayed most clearly in the inhabitant's manipulation of mathematical relations, their programming in code for computer analysis - a business of reduction and analysis within a closed symbolic and ideological domain.

The languages of different object worlds are different; their proper languages are different. In another world apart from the structural engineer, the electronics engineer speaks not of stress and strain but of power, voltages and currents, analogue and digital, resistance and capacitance. The mathematics may appear similar – there are strict analogies that apply in some instances – but the world of electronics is different, populated by different variables, time scales, units, scientific law and principles of operation. So too, different kinds of heuristics, metaphor, norms and knowledge as codified, tacit and know-how.

Participants within object worlds function as elites. But the case is different from that pictured by the philosopher Hillary Putnam.[9] Object worlds divide the design task into different, but not independent, kinds of effort so one can say that there is a "division of linguistic labour" but the distinction is not that there is one group, *an elite*, that knows the full meaning, has a god's eye view, of the object of design and another group with but a less sophisticated, common understanding of the design task. Rather there are multiple elites, each with their own proper language. It is in this sense that different participants within different object worlds with different competencies, responsibilities and interests speak different languages. Crudely put, one speaks structures, another electronics, another manufacturing processes, still another marketing, etc.

The elements of an object world language are more than words, more than the symbols and tokens of a proper language displayed by a particular scientific paradigm. I have already pointed to specialized instruments, prototypical bits of hardware, tools, ways of graphically representing states and processes as ingredients of object worlds. These all can be considered linguistic elements for that is how they function.

A sketch or more formal drawing is part of the language of design. A sketch, like a word or statement, say of an electrical circuit, can have multifaceted meaning depending upon context and intentions. Elements of a sketch may be taken from a lexicon and there are rules, regulative and constitutive, to be abided by in the drawing if the statement is to bear the author's intended sense and meaning; but meaning is far from exhausted by these rules and iconic features alone. A sketch, in what it leaves out as much as what is included, conveys notions of the object's function as well as constitution and signals both what is essential and what can be neglected.

There are other kinds of artifacts constructed in designing that may be considered linguistic. Consider, for example, a model of a vehicle crafted for use in a wind tunnel test: A physical model of this sort is as much a part of the analysis of the performance of the vehicle as is the mathematical/symbolic representation done in the proper language of aerodynamics. Its meaning is fixed in good measure by mathematical representation but not fully exploited and revealed until actually placed in a wind tunnel and put to the test. Its implications for setting specifications of the full blown object, the vehicle in the flesh so to speak, are direct. There is a mapping of the quantitative results of the test onto the big world vehicle. But I emphasize that specifications that follow from this conversation are not comprehensive for they speak only to certain properties of the vehicle, e.g., its frontal shape, the limits of dimensions of appendages and the like. For the aerodynamicist's dimensions are of a different type than those of the structural engineer. The model is not simply a geometrically scaled down version of the proposed object: Physical parameters such as the viscosity and density of air and the velocity of the anticipated airstream through which the vehicle moves enter, along with geometrical measures, into the scaling. The wind tunnel model will appear something like the big one but close inspection reveals it is distorted. We might say it looks like English but it isn't, or that the aerodynamicists sees the object of design in a peculiar way, in the light of a particular projection[10].

The icons that refer to the elements of an electrical circuit, deployed within the object world of the participant responsible for the design of the electronics, are more easily recognized as different from the physical elements themselves. Still, while common folk may recognize the wiggly line as referring to a resistor, its meaning in use is only known when made part of a circuit displayed in a sketch. Only then can one begin to think about whether a 5% resistor will suffice, what power rating should be specified, answer the question whether it provides the right RC time constant, whether it can be packaged in an array, or even what unit cost is tolerable. The tokens of object worlds point and refer but their full meaning is only constructed and revealed in the context of an object world narrative.

It is not that the persons of one world are not familiar to a degree with the language of another's world. The practitioner of any one of these worlds might have studied the content of another as part of their schooling. But studying a language in school is one thing; living and managing in the foreign country is another. Differences in context - material and conceptual - methods and instruments, codes and rules, webs of practices matter.

It should be clear from the way these linguistic elements have been described that they do not simply refer to the object in a static sense - by which I mean depicting different aspects, states, *structure* and properties of some fixed and final product - but rather are meant to describe and explain the *function* of the object of design, how it might work, when it might not, and/or the processes required to develop and produce it. They can be viewed as attempts to capture in material form, in a picture, a model, or a prototype, the *counter factual* nature of designing. Talk around, over and about them takes the form: "If we alter the airfoil shape in this manner, then the drag will be reduced by this percentage"; "If we go with the 5% resistors, our unit costs will drop by half a percent". Even a computational algorithm for Multiobjective Collaborative Optimization can be an artifact which facilitates social choice and exchange when put on the table and contested, reworked, tuned to the satisfaction of those responsible for local, disciplinary effort.

The construction and use of these varied artifacts enables negotiations among engineers designing. The things themselves are transient; varied in form, and in the process of design, in the hectic, energetic give and take, decision making and iteration, negotiation and trade off, they are active elements of a living language - shaped, specialized, reformed, extended, provoking new thought, confirming conjecture. I quote Searle.[11]

> "The unit of linguistic communication is not, as has generally been supposed, the symbol, word or sentence, but rather the *production* or issuance of the symbol or word or sentence..."

Bridging object worlds

Given this "tower of babel" vision of design process, one might wonder how it succeeds. (We note that it doesn't always do so). Well, there are methods for bringing the proposals and preferences, claims and requirements of participants of different object worlds into coherence. We have already shown two prescriptions. But how to handle the disparities I have described?

One way, a way that stands as recommended practice in the organization of any design effort, of any complexity whatsoever, is to first sit down and try to break up the task into a set of subtasks which might be independently pursued. Usually this will be done in terms of different functions the object of design must perform. Once these subsystems and subtasks have been defined and lines are drawn around them, certain "interface requirements" must be constructed and adhered to by individuals working in any two different domains. If such independently pursued tasks can be established, then participants would hardly have occasion to meet together save at some final step at which point the design would be assembled.

I claim that this is generally impossible. Not that one should not try to go as far as one can in this direction, but rather because the specification of any property, the setting of any design parameter, may be of interest to participants of different object worlds, defining interface requirements is a real challenge. Indeed, where do you draw the boundaries in the first place? And how should these boundaries be conceived? In terms of function or in terms of morphology?

In some design tasks the intensity of interaction among different object worlds might be minimal, e.g., for a product which is a re-design of last year's model, last year's organization will serve and object world language differences matter less. A tested pattern for interaction exists and provides a framework for interaction. But for truly innovative projects, e.g., the first products of a start-up, uncertainties abound and where to set boundaries, how to break up the task, is problematic. In this case, one can not foresee all of the interactions that will be required among participants working within different worlds, now organized around subtasks. One observes in this case that interface requirements are themselves subject to redesign and negotiation as design proceeds.

Granted this, we might still look for some strictly rational, instrumental methods for reconciling the differences of participants, a sort of over-arching, object world proper language to employ to our benefit. We have already seen one illustration of how this might be achieved. The scheme of Tappeta and Renaud for Multi objective Collaborative Optimization is meant to, not simply reconcile and harmonize the requirements of the different disciplines (I would say "of the participants") but to achieve an optimum resolution of their conflicting preferences.

Let us consider how this is accomplished. Each discipline has its own design objective - the f_i which are a function of some subset of the design variables - some of which are shared. To reconcile these in an optimum way, a global objective, a "...system objective function" is defined as

$$F(x^o) = \sum_{i=1}^{n} w_i \cdot f_i$$

where the w_i are some numbers, "weights", which express the relative significance of the requirements of the different n disciplines[12]:

> These design objectives are often conflicting and an assessment of the relative importance is needed for the multiobjective formulation.... In this paper it is assumed that the relative importance of each discipline

design objective is established *a priori* by assigning weights (w_i). This weighted method transforms the multiobjective function to a single system level objective function. (emphasis mine)

Now, where do these weightings come from? And why do they use the phrase *a priori?* This latin phrase means something special to, not just philosophers, but most of the rest of us. Why don't they just say "established beforehand"? But then, who does the assigning? God? The project manager? The customer, client or user? The use of the passive voice leaves us with little to go on.

Here, no doubt, is where the social intrudes. Deciding upon values for the weights will most likely require negotiation among the different participants responsible for different tasks. One might imagine a dictator making the choice but upon what knowledge base might such an omniscient agent derive his or her authority – not to speak of how such a strategy would violate most modern, enlightened managerial norms?

My claim is that the problem of bringing into coherence, none the less optimizing, the requirements and objectives (or needs) of different participants from different object worlds always exists at some level, somewhere within the design process. Setting a boundary to that process close-in and intoning "a priori" may free one from the messy business of social choice but one ought not then pretend that in this way one can dismiss the problem of harmonizing the interests of different participants by instrumental means alone.

Another way to make the claim is to say that direct translation among the different proper languages of object worlds is not possible. Comparison of the propositions and requirements of different participants requires the use of a more common or vulgar language. Here now the artifacts we have characterized as elements of language continue to function as such: Sketches made by an individual in his or her own private discourse will be dragged out for all to see, serving now as a cruder framework for the person's explanation and proposals. Whereas before they prompted detailed and exact knowledge, now they, in their new found ambiguity, provide an arena for sometimes heated, sometimes creative, deliberation and decision-making.

Claiming direct translation is not possible is akin to claiming object worlds are incommensurable. This is correct, if we take a disciplined, narrow view of the substance of such places, i.e., if we restrict our attention to the particular mathematical theories and abstract models, the variables in whose terms they are expressed, the special methods and instrumentation, codified protocols for putting their peculiar artifacts to the test, etc. The stress at the root of a cantilever beam is of another world than the open circuit voltage of a photovoltaic module. But if we stand above the fray of negotiations and collaboration, and take a broader view, we find common ground – a system of shared beliefs about how the world works, what makes it go around, whether your world or mine.

There is the mutual trust in abstraction itself. While the model of the electronic behavior of a photovoltaic cell is worlds apart from the model of the stress distribution in a cantilever beam, both the electronic engineer and the structural engineer trust in the efficacy of their respective abstractions to adequately depict what truly is essential. There is common ground in how they express the ways things work within their respective domains. Both rely heavily upon mathematical, symbolic expression, though one may make heavy use of partial differential equations, the other Boolean logic. And while their sketches and diagrams are of different form, their lines will be precise, their circles and boxes closed, their annotations cryptic but clear - in contrast to the productions of architects at a similar stage in design.

Both share the same belief in cause leading to effect: Given circumstances *A*, event *B* will follow. But, more specifically, both claim to be able to quantify, to measure the relevant ingredients of *A* and those of *B*. If the solar flux onto the module is 1 kilowatt per square meter and the ambient temperature is 20 degrees centigrade, then the maximum power available out of the module will be 60 watts. If the weight at the end of the cantilever is 100 pounds, and the beam itself weights 20 pounds, then the maximum stress at the root of the beam will be 4000 pounds per square inch.

Both can demand of the other verification of proposals and claims via the testing of hardware and prototypes constructed in accord with the concepts, principles and purposes of their respective worlds. Each will accept the other's *ceteris paribus* stipulations, explicit or implied, as these are part and parcel of the conceptual schemes of different and independent object worlds.

Both take a strictly instrumental view of their productions – of knowledge, of function. Simplicity is valued. Being in control is valued. The two go together. Technical perfection, e.g., optimization, is possible within object worlds. Color is generally irrelevant; aesthetics is secondary; even costs garner little respect - although of course they have to be dealt with. So too codes and regulations, marketing directives, lawyers warnings, the CEO's proclamations – these are all ingredients of design but life within object worlds can go on without them, indeed, much more neatly without them.

There, in more than a nutshell, is my way of seeing engineering design/designing engineers. Different participants with different responsibilities, competencies and interests, speak different languages when working, for the most part alone, in their respective domains. For this to ring true, we ought to construe language in the broadest terms - to include the sketch, the prototype, the charts even a computer algorithm as elements employed in the productive exchange among participants. But individual effort within some disciplinary matrix does not suffice: Designing is a social process; it requires exchange and negotiation as well as intense work within object worlds.

Notes.

1. Bucciarelli, L.L., *Designing Engineers* (Cambridge, MA, MIT Press) 1994.

2. Bucciarelli, L.L., "Power Loss in Photovoltaic Arrays due to Mismatch in Cell Characteristics," *Solar Energy 23* (1979): 277-288.

3. Google: Advanced Search, Exact Phrase "the world of", February, 2002.

4. Dixon, J., *Design Engineering*, New York: McGraw-Hill, 1966.

5. Pahl, G., and Beitz, W., *Engineering Design: A Systematic Approach*, (trans. K. Wallace, L. Blessing, and F. Bevert), 2nd ed., Springer 1996, p. 24.

6. Tappeta, R.V., and Renaud, J.E., "Multiobjective Collaborative Optimization", *ASME Journal of Mechanical Design*, September, 1997, vol. 119, pp. 403-411.

7. A good bit of this section has been adapted from Bucciarelli, L.L., "Between Thought and Object in Engineering Design", *Design Studies*, 23, (3), pp. 219-232, copyright May 2002, and is included with permission from Elsevier Science.

8. Ryle, G. "Knowing How and Knowing That", *Proceedings of the Aristotelian Society*, vol. XLVI, 1946.

9. Putnam, H., "The Meaning of 'Meaning'", in *Mind, Language and Reality*, Cambridge, 1975.

10. Wittgenstein, L., *Philosophical Investigations*, Prentice Hall, Englewood Cliffs, NJ, 1958.

11. Searle, J., *Speech Acts: An Essay in the Philosophy of Language*, Cambridge University Press, Cambridge, UK, 1969, p. 16.

12. Tappeta, et al, op. cit., p. 404.

What engineers don't know & why they believe it.

That engineers may *not* know something, or that they don't know something, will be generally accepted as fact, though the impression given by some might be otherwise. But it may strike you as presumptuous, that I, or anyone else for that matter, might be able to describe what it is they do *not* know. More problematic, I want my essay to have meaning to engineers themselves so that, while admittedly, there is much that engineers don't know - e.g., how to juggle, the words to the national anthem of the Netherlands, who won the world cup in 1986 - our concern is with that which they don't know which is relevant to their day-to-day existence and professional practice - that which they *should or ought* to know or would be better off if they *did* know.

The chapter title as a whole might strike the reader as a bit of a muddle. I could straighten it out by adding a qualification, to wit: "What engineers don't know *with certainty but* (and) why they believe it to be the case". But this weakens the claim too much; it rules out important cases of "not knowing", e.g., the possibility that the "it" can be that which is unknown, un-thought-of, unmentioned, unseen, unheard, unfelt, unimaginable. I don't want to rule out this possibility; hence I will not amend my title[1].

Knowledge presumes belief; belief rests upon trust; trust is a social matter; trust binds beliefs and people together, pervading the different contexts within which the engineer must function. Within the context of the design task itself, there is the trust among participants of different object worlds. Within any object world one trusts in the integrity of the dictates and heuristics of the defining technical paradigm. Within the context of a supporting infrastructure, engineers rely upon the claims and promises of suppliers, subcontractors, parts manufacturers. And users and customers are trusted to behave, to respond as imagined and specified.

Introducing "trust" as an essential aspect of engineering work is one of a piece with viewing what engineers do as social as well as scientific. The challenge of this enquiry is to relate engineering thought and practice seen as a social, as well as an instrumental process, to engineering thought and practice seen as a subject for philosophical critique and analysis. I try to straddle both worlds, of social study and philosophy, holding as I do that by entertaining both perspectives we can construct a better understanding of what engineers know, what they believe, what they do, and how they might do better.

We begin by noting that within all design contexts there are uncertainties. Some of these may be identified explicitly, given probabilistic expression and thereby brought within an object world for instrumental assessment. But there remains the possibility that what we believe will not prove to be the case. Whatever the context, there are (yet to be) relevant things engineers don't know, and yet they believe and trust they are in control.

The suggestion that engineers don't know as much about the integrity of their productions as they believe to be the case is prompted by the simple observation encapsulated in 'Murphy's Law' i.e., that if things *can*, things *do* go wrong. Products, processes, systems fail. In this chapter, I want to explore the nature of technical failure and how engineers and other participants in the technological enterprise cope with malfunction and attempt to set things right. I am interested in whether philosophy is relevant in any way to developing a better understanding of this kind of engineering practice, making clear, on the one hand, what might be changed to improve practice and what, on the other hand, must necessarily remain problematic.

The Nature of Technical Failure

What constitutes mis-behavior, a failure event? Some are obvious: The Hyatt Regency walkway fails dramatically under the load it was designed for causing the loss of life and limb. The Tacoma Narrows bridge oscillates so wildly in the wind, it collapses. But other events, most perhaps, are not so easy to identify as failure, none the less uncover their cause. The "yield" of an industrial process, say for the production of silicon wafers destined to be computer chips, is not as high as desired *some* claim. Others argue that the yield is good enough - and besides, the cost of improvement of the production process is not justified. So "failure", malfunction, can be a matter of degree. The software application when run within a particular operating system hides a dialogue box when I return to the main window displayed on my monitor. Is this a "bug" or a "feature"?

So whether misbehavior is deemed significant or even to be defined as such depends upon who you ask. Malfunction can be described and defined with respect to a set of performance specifications; with respect to the expectations of participants in design; or with respect to the expectations of users. These are different, more or less independent referents: A product or system may fail to meet a specification yet satisfy the customer. Alternatively a product may meet all specifications yet not satisfy a participant in design and may, or may not satisfy all or any users. And, of course, a product may meet specifications and designers expectations yet fail in the marketplace.[2]

Failure is related to the quality of a product - the other side of the coin so to speak. And just as the quality is difficult to define (good to, or for whom?) so too what is seen as failure is not a wholly objective matter. For the purposes of this essay, *I will take failure of a technical production as that event which engenders, or would likely engender, corrective action on the part of those responsible for its design, making, operation and/or maintenance.*

In this sense, failure is a social construct. That is, whether an event is labeled a failure depends upon the beliefs, judgements and claims of persons concerned with the event – claims which are taken seriously by those responsible for the design, making, operation and maintenance of the product or system who themselves, of course, are persons so concerned. These "participants" in the construction of failure may be a varied lot, each with a different perspective on the product's nature, its function and use. As in design, where different individuals see the same object of design differently in accord with their technical competencies, responsibilities and interests, so too in the construction and analysis of error, different concerned parties will see the failure event and its ingredients differently. Defining failure is a social process.[3]

Diagnosing failure

Once a malfunctioning has been defined, the search for a cause begins. Finding the cause, one can then try to fix matters so it doesn't happen again. Diagnosing technical failure is not too different from any effort which, when confronted with the symptoms of illness, one strives to move beyond the appearances to expose the source and reasons for the malfunction. For this, in engineering as in medicine, there are certain strategies.

When things go wrong, when your product or system mis-behaves and surprises you or, more ambiguously only suggests that something is out of order, the first task is to try to replicate the failure, to establish conditions such that, when we set the system in motion, the faulty behavior re-appears.[4] In this way we can construct a fuller description of the problem, stating under what conditions and with what settings the fault occurs.

A next step, if we are successful in replicating the malfunction - and this with some consistency - is to change conditions in some way and observe the result. We seek to make a relevant difference in conditions - relevant in that it alters the state of the product in some significant way - and then note if this alteration does, or does not, eliminate the failure. Our traditional strategy recommends that we change but one condition at a time, proceeding in this way until a cause of the failure is identified, i.e., the system stands corrected and now runs as it should.

In this process, we ordinarily have more than the product alone to work with. Drawing upon our knowledge of how the artifact was designed – i.e., in accord with certain scientific concepts, principles and instrumental methods– we can construct a mathematical representation or physical scale model of it's behavior and put this to the test. Indeed, we most likely already have these alternatives available since they are essential to designing in the first place.

Sometimes a model may be the only feasible way to test alternative scenarios; the real artifact may be inaccessible for one reason or another. Still, this may suffice: For, with the model in hand, we can alter inputs and/or parameter settings and observe the result seeking, as we would if it were the actual product, to provoke the faulty behavior.

This was the challenge faced by engineers at NASA when a solar panel on the Mars Global Surveyor failed to deploy.[5] The Global Surveyor, launched in November of 1996, was designed to travel from Earth to Mars, there go into a circular orbit and collect data about the geological nature of the planet. As the insert reports, one of the two solar panels evidently failed to fully deploy shortly after launch. The primary concern expressed in this news release was the effect of the failure on accomplishing the mission's objectives. At the time, engineers concluded that the skewed panel would not significantly impair the spacecraft's performance. (Perhaps this is why the failure is not labeled as such but referred to as "the situation"). This conclusion was reached using computer-simulated models and engineering tests - the latter referring to tests of duplicate hardware components designed to drive and control the deployment of the solar panels - for the failed artifact was not available.

What was available was a stream of data sent back from the spacecraft i.e., "...two weeks of spacecraft telemetry and Global Surveyor's picture-perfect performance during the first trajectory maneuver...". That the panel had not fully deployed might have been made evident if the craft had been instrumented to signal ground control when latched into the desired final position. That the panel was shy by 20 degrees might have been inferred from the electrical output of the photovoltaic array as a whole. Just what the "picture-perfect" performance had to do with the diagnosis is not clear but the phrase tends to lead one to attribute too much to what they did "see". For it was the ground-base computer models and engineering tests, together with the telemetry data alone, which provided a basis for an explanation of why the panel had failed to deploy fully, not any photographic, video, or film image. At any rate, though they could not see the spacecraft, they had sufficient reason to believe one panel had not fully deployed. What caused the damper arm to break is not explained nor conjectured.

FOR IMMEDIATE RELEASE   November 27, 1996

GLOBAL SURVEYOR SOLAR PANEL WILL NOT HINDER MISSION GOALS

Mission engineers studying a solar array on NASA's Mars Global Surveyor that did not fully deploy during the spacecraft's first day in space have concluded that the situation will not significantly impair Surveyor's ability to aerobrake into its mapping orbit or affect its performance during the cruise and science portions of the mission.

The solar panel under analysis is one of two 3.5-meter (11- foot) wings that were unfolded shortly after the Nov. 7 launch are used to power Global Surveyor. Currently the so-called -Y array is tilted 20.5 degrees away from its fully deployed and latched position.

"After extensive investigation with our industrial partner, Lockheed Martin Astronautics, using a variety of computer- simulated models and engineering tests, we believe the tilted array poses no extreme threat to the mission," said Glenn Cunningham, Mars Global Surveyor project manager at NASA's Jet Propulsion Laboratory. "We plan to carry out some activities in the next couple of months using the spacecraft's electrically driven solar array positioning actuators to try to gently manipulate the array so that it drops into place. Even if we are not able to fully deploy the array, we can orient it during aerobraking so that the panel will not be a significant problem."

Diagnosis of the solar array position emerged from two weeks of spacecraft telemetry and Global Surveyor's picture-perfect performance during the first trajectory maneuver, which was conducted on Nov. 21. The 43-second burn achieved a change in spacecraft velocity of about 27 meters per second (60 miles per hour), just as expected. The burn was performed to move the spacecraft on a track more directly aimed toward Mars, since it was launched at a slight angle to prevent its Delta third-stage booster from following a trajectory that would collide with the planet.

Both the telemetry data and ground-based computer models indicate that a piece of metal called the "damper arm," which is part of the solar array deployment mechanism at the joint where the entire panel is attached to the spacecraft, probably broke during the panel's initial rotation and was trapped in the 2-inch space between the shoulder joint and the edge of the solar panel, Cunningham said.

Engineers at JPL and Lockheed Martin Astronautics, Denver, CO, are working to develop a process to clear the obstruction by gently moving the solar panel. The damper arm connects the panel to a device called the "rate damper," which functions in much the same way as the hydraulic closer on a screen door acts to limit the speed at which the door closes. In Surveyor's case, the rate damper was used to slow the motion of the solar panel as it unfolded from its stowed position.

When restricted in this way to the use of an abstract representation alone, we must ask if the model is good enough, e.g., conceptually adequate, complete, and sufficiently accurate. In some cases we are able to "repeat" the mis-behavior with the model as is, without changing the formal structure of the latter; i.e., we change an input condition or parameter and replicate the malfunction. In other cases we find we must change the structure of the model. In this case we will fault the model as well as the artifact. (The design was in accord with the model). In either case we take action and then claim a fix, believing and trusting in our new picture. But how do we know the model, even if updated, is an adequate representation of the real thing? Might not we be missing some relevant detail? Might not there be other conditions which would result in the same faulty behavior?[6]

Indeed, as the Global Surveyor entered the outer fringes of the Mars atmosphere and the "aerobraking phase" of the mission began - a maneuver intended to bring the craft into the desired, final circular orbit about the planet - NASA engineer's picture of the failure had to be repainted: The braking force on the unlatched panel moved the panel past its latched position. This meant that something more than a broken damper arm was involved as a "cause" of failure (else why wouldn't the panel have latched)? Engineers now conjectured that a supporting structure at the interface where the deployment mechanism met the body of the spacecraft had failed. The immediate fix, if it can be called that, included performing the aerobraking maneuvers at a slower rate which, though changing the craft's final orbit about Mars, would not detract from meeting the mission's objectives.

Even, in general, if the actual artifact is available for diagnostics, the strategy sketched out above has lacunae. Consider the first step - replicating the failure. Even if successful in this, we are not justified nor have any sure basis in claiming that the set of conditions that precipitates the malfunction observed is the *only* set of conditions that might result in the same symptoms, the *same* failure; nor, for that matter, are we justified in claiming that the next time we set these *same* conditions the failure will be made evident. For, in the latter case, if there remains some condition that is not under our purview but is relevant and, conjoined with others on our list, alters the state of the world such that the machine works as it should, we will be pleasantly surprised, or rather frustrated, in this outcome. We can never be sure that "all other things remain equal".

Underwriting this critique is the claim that a design is *under-determined* in the sense that all possible 'behaviors' (i.e., functioning, workings, states, input-output response) are never fully determined or forecast in the course of the design process.[7] There are at least three sources for under-determination: Within object worlds there are limits to predictability due, in part, to lack of resources e.g., time, or inability to fully replicate the context of use. Then too, some 'inputs' are difficult to capture in an analytical mode, e.g., parameters which are difficult to quantify, their range uncertain. How does one model the quality of maintenance or the possibility of an antagonistic, or even ill-intended user? While some uncertainties might be dealt with probabilistically, there are still other features which remain unknown.

A more problematic source of under-determinacy lies in the unanticipated interaction among the design contributions of participants from different object worlds. It is difficult to predict all the interactions across the interfaces established to enable participants to work as independently as possible. Nor can the significant details of all contexts of use be anticipated. Analytical exactness and completeness may hold within object worlds but the behavior of the whole, in a sense, is not fully defined by the behavior of its parts. It is this fundamental feature of designing which both makes engineering the challenge that it is and denies the possibility of achieving technical perfection.

It also reveals the naivete of viewing of engineering design as the straight forward, rational application of science. There is indeed a resonance here with science - in the philosophical position that holds that one can never fully verify a scientific theory. But note a contrast with the scientific enterprise: With the latter, one strives to deduce all significant consequences of the theory. In engineering, one strives to ensure that the product meets some limited and prescribed specifications whatever else it might do, or be encouraged to do.

Under-determinacy insures as well that the challenge of diagnosing error (or anomaly) will require more than the application of rational, instrumental method. While engineers may believe and trust in the integrity of their productions made in accord with the dictates of object world instrumental thought and practice – having no reason to judge otherwise – there still remains the possibility, nay probability of things going wrong once the product is launched out into the world.

A scenario: Failure of a truss structure.

Here I construct a simple example of the complexities of diagnosing failure. It is a thought experiment; my intent is to bridge the disconnect between philosophy and engineering by adopting a well known scenario of philosophers which concerns criteria for knowledge claims. As such it is story of what might in fact occur - a scenario within a possible world. No where do I violate the second law of thermodynamics. My report is in complete accord with the dictates of object world thought and practice in structural engineering.

The figure shows a bracket meant to support a weight at point *C*. This could be a shelf bracket, for example, or support for a sign, or even a critical substructure of a weapon's system – no matter.

The arrangement is commonly known as a "truss structure". The arrows indicate the directions of the force exerted on the point *C* by the weight *W* and of the vertical component of the displacement, *d*, of the point *C*. The magnitude of the displacement depends upon the magnitude of the load, the material out of which the truss members are made, their lengths and cross-sectional areas. There is a theory, that of "elasticity" or more immediately, that of the "strength of materials", which enables a structural engineer to predict the magnitude of displacement given this information.

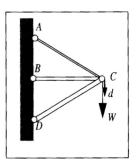

Figure 3.1 A truss structure

In my scenario, as well as in general, the system meets specifications if, when loaded with the anticipated weight, *W*, the structure retains its shape - there is but a small, mostly vertical displacement of the point *C* when loaded. Of course, it is possible that the tension or compression in any one of the three members might become excessive - say as the load *W* was increased without limit - and a member breaks, fractures, or deforms excessively like a soft plastic. This would certainly be considered failure.

In my particular scenario, it was specified that the displacement *d* due to the anticipated load *W* ought not to exceed a certain limit, i.e., the structure was to exhibit a certain minimum "stiffness". The properties of the three steel members - their cross-sectional areas - were chosen to ensure that the displacement remained within the prescribed bounds when the maximum value of the anticipated load was applied.

When put to the test, however, the structure proved more flexible than desired. At the anticipated load levels, the displacement of point *C* was excessive. Hence, the structure did not meet specification, though the members did not break at the design loading.

Confronted with this deficiency, a fix was proposed and accepted. The bottom member, *DC*, was replaced by a stiffer member, one with 20% more cross-sectional area, so that the overall stiffness of the structure was likewise increased by a comparable percentage. Tests showed the new structure met specifications; the displacement at the anticipated load was no longer considered excessive. All was put back in order; the structure could be released for use. It was believed it would now behave in accord with specifications.

Unfortunately, when installed in the field, the bracket was fastened upside-down to the wall; i.e., member *DC*, the stiff member was located on top, *AC*, the not-so-stiff member, positioned and fastened to the wall at the bottom. When the structure was loaded with the weight *W*, the displacement observed, when the magnitude of *W* reached the design load, was again excessive and, what was worse, appeared to be increasing at an alarming rate with each small increment in the value of the load.

Upon closer inspection, it was discovered that the member *AC*, now in compression, had experienced major deviation from a straight line; it had "buckled". This explained why a small increase in load in the vicinity of the allowable load level engendered very large additional vertical displacement of the node *C*.

Why did it fail? If the truss had been installed "correctly", i.e., with member *AC* located on top, it would not have failed; the laboratory conditions would have prevailed in the field. But this was not the case; the context changed; no one considered the possibility that the structure would be installed upside down.

To be "unthought-of", not considered, is to remain "unknown". The claim here is that there will always be a potentially problematic state of affairs not considered, overlooked, unimagined, unconstructed, no matter how many safety procedures one invokes or how imaginative and free wheeling your brainstorming session about possible contexts of use may be.

Oh, but you say: "Your scenario has a point but any engineer worth his or her salt would display better design practice. Surely one would have tested the truss in both configurations and stiffened up the structure in other ways to make it accord with specifications - no doubt both members should be replaced with identical, stiffer members."

No doubt? Surely this could not happen? What kind of a response is that? Of course I am allowed my doubts. What justification, how can one be so sure? Where have I erred in my scenario?

Another might chime in: "I would design it so that it would be installed in the proper configuration - the simplest thing would be to add a label, i.e., 'this end up'; or better yet, design it so that it can only be installed in one way, the correct way. Do this by making the angle that the top member makes with the horizontal, less than that of the bottom member. Then since the wall sockets are in place, there is only one way the truss could possibly be installed".

The latter's advice was taken, and member *AC* then met the horizontal at a shallow angle. The system was again deployed. But once again the behavior was like before - excessive displacement at the design load and evidence of non-linear increase in displacement with load. This time, however, inspection showed that the bottom member had not buckled; rather the top member AC had deformed dramatically in tension – much more than it should have. How could this be? After all, the members had been chosen to support the anticipated load.

A "root cause analysis" conducted with all participants in assembly as well as design and manufacturing revealed that member *AC* had been fabricated a bit short. When installed it had to be stretched, pretensioned to connect up to support point A at the wall. This pretension, together with the additional tension engendered when the load was applied at *C*, exceeded the yield stress of the material.

And so it can go...

While whatever fix *I* *make* within the scenario eliminates from the realm of possibility one more failure mode; whatever additional recommendation *you* *make* for improving the design process, e.g., let's improve quality control; whatever redundancies you might add to the system to take care of whatever odd circumstances of the context of use you are able to forecast; whatever retreat to probabilistic construction of acceptable risk of failure you make; I can always imagine a new state of affairs, conditions within a possible world - in designing, in manufacturing, in assembly, in packaging, in use, in maintenance - which would be unaccounted for, unthought of, and which would engender failure (in the mind of someone).

This scenario is not so much a fantasy as suggested: In attempts to ensure safety-in-use of a new product, participants in design will themselves play out scenarios of possibilities which might endanger the user. The claim here is that the set of possibilities will never be complete; there will remain the possibility that some "idiot" will, against all expectations, do something that will endanger his welfare or that of society.[8] My scenario is about a simple truss structure; it is chosen as simple as possible to show that even in such cases the possible existence of unknowns leads to the conclusion that one can never fully verify a design. One need not consider so called "complex systems" to make the point.

I now alter this story to make an explicit connection with philosophy: This revised scenario is meant to contrast with one advanced by Gettier.[9]

In his article, Gettier challenges his contemporaries attempts to state necessary and sufficient conditions for someone knowing a proposition, p, is true. One such set of conditions has the form:

s knows that *p*   if and only if

(i) *p* is true

(ii) *s* believes that *p*, and

(iii) *s* is justified in believing that *p*.

Gettier tells the following story which shows that these conditions are not sufficient to enable one to claim that "s knows that p".

Smith and Jones have applied for a certain job. And suppose that Smith has strong evidence for the following conjunctive proposition:

(d) Jones is the man who will get the job, and Jones has ten coins in his pocket.

Smiths's evidence for (d) might be that the president of the company assured him that Jones would in the end be selected, and that he, Smith, had counted the coins in Jones's pocket ten minutes ago.

Proposition (d) entails:

(e) The man who will get the job has ten coins in his pocket.

Let us suppose that Smith sees the entailment from (d) to (e) and accepts (e) on the grounds of (d), for which he has strong evidence. In this case, Smith is clearly justified in believing that (e) is true.

But imagine further that unknown to Smith, he himself, not Jones, will get the job. And, also, unknown to Smith, he himself has ten coins in his pocket. Proposition (e) is then true, though proposition (d), from which Smith inferred (e), is false. In our example then, all of the following are true:

(i) (e) is true

(ii) Smith believes that (e) is true

(iii) Smith is justified in believing that (e) is true.

But it is equally clear that Smith does not know that (e) is true; for (e) is true in virtue of the number of coins in Smith's pocket, while Smith does not know how many coins are in Smith's pocket, and bases his belief in (e) on a count of the coins in Jones's pocket, whom he falsely believes to be the man who will get the job.

Gettier, E.L., "Is Justified True Belief Knowledge?", Analysis, 23.6, June 1963 pp 121-123.

I illustrate, as Gettier did, that this set does not work:

Suppose Jane is a new hire recently charged with indeterminate truss customer relations. She arrives on the scene after the first fix was made - her predecessor had signed-off on increasing the stiffness of the bottom member, *DC*, but reports were coming in that this, in some cases, had not solved the problem. Last week, there was a report of a wrong-headed installation and a buckled bottom member.

This week, she gets a call from a customer who complains of excessive displacement under a specified, allowable load. Jane goes to observe on site the performance of the truss. Unfortunately, the customer has encased the structure within a housing which can not be easily removed. Jane has to rest content with observing only how the node C displaces as the load W is increased. (Here note that, though the artifact is available, it is not available to the full extent that Jane would like). She takes the data back to the office and indeed confirms that the displacement is excessive and appears even to be a bit non-linear. In the lab, she sets up the truss "upside-down", loads point C with the allowable weight W, and observes the onset of buckling.

Jane has strong evidence for the following conjunctive proposition:

(d) the truss was installed upside down and the top, now bottom, member had buckled.

Jane's evidence for (d) might be the precedent cited above and Jane's own test of the truss, configured upside down in the lab - which did indeed produce onset of buckling at the design load.

She posits:

(e) Increasing the cross-sectional area of the top (AC) member will rectify the problem.

But imagine further that, unknown to Jane, when the protective housing was installed the truss was set in the right orientation but an interference was encountered at the top junction where the member would normally be fastened to the wall and that the top member (AC) was subsequently placed under considerable pre-stress when the force fit to the wall was made. Under the design load, plastic deformation occurred in the top member which in turn caused the excessive vertical displacement at C. Despite this, carrying through the fix in accord with (e) will solve the problem.

Proposition (e) is then true, though proposition (d), from which Jane inferred (e), is false. In our example then, all of the following are true:

(i) (e) is true

(ii) Jane believes that (e) is true

(iii) Jane is justified in believing that (e) is true.

But it is equally clear that Jane does not know that (e) is true; for (e) is true in virtue of the plastic deformation of the top member, AC, while Jane does not know of the plastic deformation of the top member and bases her belief in (e) on the buckling of the bottom member which she falsely believes to be the cause of the failure.

This Gettier type counter-example to the three conditions for what might count as knowledge, patterned on one of the scenario's found in Gettier's own article (see the boxed text), is one in which Jane has a justified but false belief, i.e., (d), by inference from which she justifiably believes something which happens to be true, (e), and so arrives at a justified true belief which is not knowledge.[10]

Now I am not going to pursue how philosophers have contended with Gettier's counter-example, have tried to amend the conditions for knowledge so that the counter-example looses its force or built upon his provocation to redefine conditions for knowledge. But I do find something provocative in his challenge, something that I think has relevance to the task of ensuring the integrity of our technical productions, or at least acknowledging the limits in design. As my adoption of Gettier's scenario suggests, whether or not we have a hold on the true cause of misbehavior might not prevent our making a legitimate fix. The problem with this is that this "false knowledge" - if I can call it that - while it may serve our immediate purpose, is very likely to bring us trouble in the future.

If I were to offer a recommendation, it would be that the engineer might do well to adopt the perspective and attitude of the skeptic when challenged with the task of diagnosing failure. While there are rules and heuristics, block diagrams and instrumental, rational methods for dealing with a malfunctioning production, one can never know for sure whether one has a solution for all times, all contexts, for all possible worlds.

A more complex world

This last scenario, together with my crude description of diagnostic practice, is abstract, theoretical and meant to make evident, even in the simplest of cases, the challenge of determining the cause of technical failure. While providing some sense of the limits to engineering knowledge and knowing, it stands apart from the world of actual practice. It is like the block diagrams of design process – those abstract and aloof prescriptions for designing displayed in my first Chapter. Recall that the latter representations are all about method; all focused on the product; there are no persons there within their boxes and borders. They imply that the design task may be the sole responsibility of a single agent - an individual, a firm, an institution - we know not which. There is nothing "social" in them. So too in Gettier's world: Jane, Smith and Jones, are hard to see as social agents. But that is irrelevant to the meaning of the story: What matters is that their exchange, the propositions and inferences made, is in accord with accepted norms of philosophical analysis.

I want now to make matters more complex through two summaries of examples of technical failure drawn from the recent past and the so-called real world. Here now we encounter participants in the diagnostic task having different responsibilities and interests and, as in the design task, the reconciliation of their different claims and conjectures requires more then instrumental analysis and object world work, more than computer modeling and hardware testing, more than analyzing telemetry data and more, for that matter, than a Jane or a Smith or a Jones who has but ten coins in his – rather, in their pockets.

Negotiating quality

My first example is drawn from Diane Bailey's "Comparison of Manufacturing Performance of Three Team Structures in Semiconductor Plants".[11] Her aim is to test the notion that workers, if allowed greater freedom in decision making when confronted with production problems on the shop floor, will put skills and knowledge to more effective use than will employees going by the book, so to speak.

In this, she lays out the results of her comparative analysis of the performance of manufacturing work groups organized in different ways. One group was organized in self-directed work teams (SWDT); other groups she studied were organized in continuous improvement teams (CIT) and in quality control programs (QC). She finds that the productivity of the self-directed work teams of etch operators is poor when compared with the other groups which were organized differently. This runs counter to her expectations: She had expected the self-directed work teams to do better. She seeks to explain their relatively poor performance:[12]

> One could argue that time spent by SDWT's... on resolving quality problems would necessarily reduce productivity as measured here by the number of wafers produced, but that overall production of good chips (from the wafers) would improve by increasing the percentage of good chips per wafer. This latter metric is referred to as die yield; unfortunately, the complexity of the production process precludes the tracking of die yield problems to individual functions, let alone workgroups.

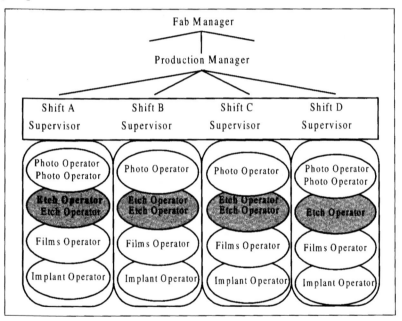

Figure 3.2 Typical fab production hierarchy

The complexity is due to the fact that a typical chip had about 16 layers, each requiring about 4 machine processes resulting in 64 possible processes to consider as culprits as the cause of any defective chip. The number of machines involved in these processes increases further the number of potential sources of defective chips.

> ... the problem primarily arose from having many suspects for each particular type of defect and many more (types of) defects than anyone had

ever catalogued up to that point. The problem was further complicated by not being able to test for most defects after each individual process; the entire circuitry needed to be in place before testing could be done.[13]

One might conjecture that using die yield – the number of good chips per wafer – rather than wafer count, as a measure of the relative effectiveness of the self-directed work teams might give a very different measure of the quality of the production process. The definition of failure of the production process then becomes problematic. For the manager of wafer production, the lower percentage of good wafers, relative to other groups, indicates a substandard process; for quality control, further along in the production process, a higher die yield might suggest the contrary. Whether there is a failure of process then becomes a matter of who you ask.

Even if there is agreement on both counts that the process needs to be improved, i.e., both measures indicate there is a problem and both parties to the process agree that corrective action must be taken, there remains the challenge of determining the cause(s) of the failure and prescribing a fix. This too will involve more than instrumental reasoning: Negotiation of interests and perspectives will be required.

In this, the different competencies and responsibilities of participants in the fault-finding process point in different directions: The manager of production looks at the process (as object) and sees one thing – too much scrap in the production of the wafers. She worries about organizing and motivating her work force to do better. The head of Quality Control looks at the process and reads the problem in another way. He sees a more detailed and complex picture of machines and circuitry, each able to stray with time. Despite the multiplicity of ways defects might be engendered, he holds that attacking the problem at this level of detail is the only way to obtain sure knowledge and be fully justified in advancing a fix. From his perspective, management appears only to be concerned with surface appearances and uninterested in getting at the real causes of the group's poor productivity. The manager of production, on the other hand, views her colleagues proposals as too costly and even unnecessary. Like Jane, when confronting the failed truss structure encased in a housing, she deems it unnecessary, none the less unfeasible, to open up the black box containing the chip processing and fabricating operations, to look inside, search for, and determine the "real" - according to Quality Control - sources of the problem.

High technology products exhibit this layered complexity. In our diagnostic activities we would like to get at the root cause of the failure. But at what level of detail do we stop? And what about the resources required to make this journey? Limits of time as well as money sooner or later press for closure, for making a patch - what some would consider a less than satisfying fix - and getting on with the business. Less pragmatically, we might ask if the idea of a "root cause" in itself – a single factor which when identified explains all and, if corrected, renders performance near perfect – is but a fantasy. While a rhetorical construction of such is always possible, we can question its ontological status, even if the consensus is broad. My next example illustrates the challenge of finding a single causal factor.

Negotiating causes

This example differs in at least two ways from the previous one: Here there will be no question about whether failure occurred. Second, the participants in the diagnostic process in this case are not all members of the same firm; control and definition of failure will rest in the hands of different and independent agents.

The failure concerns the performance of the Bridgestone/Firestone Radial ATX and Wilderness AT tires. As reported by the US National Highway Traffic Safety Administration

> Tread separation claims included in the Firestone claims database involving the recalled and focus tires have been associated with numerous crashes that have led to 74 deaths and over 350 injuries (as of March 2001).[14]

While I am not going to try to analyze in detail this tragic and costly failure, I do hope to describe who was involved in the diagnostic and remedial process, lay out what factors were construed as significant, and sketch how these were negotiated. Participants included, but were not limited to; Firestone Tire; Ford Motor Company whose Explorer SUV ran on the Firestone tire; the owners, drivers of this large vehicle; the National Highway Traffic Safety Administration and we might add to the list the hot sun on southern interstate highways.

Clearly in this case there was a failure. Belief in the connection between tread separation, primarily on the rear tires, and the crash of the Explorer was amply justified by the evidence at hand and accumulating. But when we focus on attempts to determine the cause of failure, we find, as we might expect, different participants advancing different claims and conjectures about the root cause and what should be done to set matters right.

The crashes, according to Firestone, were *not* due to one over-riding factor, i.e., high stress levels leading to tread separation, but derived from a combination of factors, e.g., low inflation pressure together with an overloaded vehicle. A question contested throughout the debate - or negotiations if you prefer - between Firestone and Ford was what tire pressure was required to ensure that the tire temperature did not rise to the point where tread separation would be likely. Firestone recommended a higher pressure, which meant a harder ride, Ford a lower pressure. Firestone also concluded from tests that the Explorer vehicle allowable load levels which were set by Ford, for the tire pressure initially specified for the Explorer, again by Ford, would approach the limits of the tire's load carrying capacity. Tires on average lose about one psi per month so in four months "...the left rear of an Explorer would be overloaded."

Ford of course, held that the root cause lay in Firestone's marginal design of the tire, coupled with poor quality control at Firestone's Decatur plant – a fact the tire manufacturer allowed though claiming the Decatur tire still met Ford's specifications.

Firestone, at one point, claimed that the shoddy design of the Explorer was the root cause. In a letter to NHSTA, asking the agency to open a safety defect investigation into the handling and control characteristics of the Ford Explorer following a tread separation on a rear tire, they claimed

> ...that the crashes (many of which involved rollovers) that occurred in Explorers following tread separations of Firestone ATX and Wilderness AT tires were to a large extent due to the design of the Explorer rather than a defect in the tires.[15]

Here, more obviously, we have a diagnostic task which includes different participants with different interests, responsibilities, competencies and perspectives. Here too a task which required object-world work: the engineering analysis of internal stresses in the ATX and AT tires and whether they exceeded allowable levels; the experimental measurement of increase in temperature of the tires after running under load at high speeds and how this depended upon the initial pressure in the "cold" tire; the dynamic stability of the Ford Explorer under normal operating conditions and when perturbed by the blow-out of a rear tire. But these studies were not, in themselves alone, definitive: Negotiation of cause and fixing of responsibility builds upon object-world knowledge but this knowledge, while necessary, is not sufficient.

While the matter may have quieted down, only recently we find a news report of another tire recall not unrelated: Continental Tire North America has recalled half a million tires installed on Ford's sport utility vehicles because some of the tires had lost their tread.[15]

Consider now, in a skeptical vein, if the Ford Explorer did not exist – a possible world we must admit. Would the failure have happened? That is, would the occasions when these same tires on other vehicles blew out and caused a crash, even death of the vehicle occupants, been of sufficient frequency so that a failure would have been identified? Would the determination of cause play out in the same way? We might even go further and ask: What if government regulations on gas millage precluded labeling the Explorer SUV as a light truck. Would there have been a Ford Explorer? Should we count government inaction, the existence of this loophole, as a contributing cause? Where do we draw the boundary? How far do we go before claiming all else remains equal (or irrelevant, or non-contributory)?

Conclusion

How do engineers cope with error? I have tried in these stories, all concerned with technical failure, to shed some light on the nature of error, how it can or cannot be explained, and what it requires to construct a remedy, "a fix".

As engineers, we stand as society's role model of rational, instrumental thinking. We would like to proceed with some assurance that we can identify the true cause of malfunction in our designs, products and systems and so be certain and confident in our proposals to set things right. This, from the perspective of the skeptic, is not possible. Though one may be justified in one's belief that a fix will hold for all time, one does not know this will be the case. Of course that does not mean that one should not strive to uncover the true cause, or causes, of failure, should not make use of whatever rational, instrumental methods are promising, but it does set a limit on what might be claimed as certain knowledge. And while a complex system may present more of a challenge, the result holds true for products of the simplest kind. There is always the possibility of some unthinkable event happening within some bizarre context - in design, in manufacturing, in use - provoking unanticipated behavior, which may or may not be construed as "failure".

This last observation points to a further complexity - the social nature of "failure". By that I mean not only that the actions of people, agencies and social institutions may be factors in failure, but that addressing and setting things right requires negotiation among different parties to the problem in order to identify, define, and resolve matters. This, then, makes diagnostics like designing: And just as a design is under-determined, so too any final construction of failure and its remedy rests under-determined.

At the root of this claim lies the possibility that different participants in the design or the diagnostic process see the world differently and that these differences matter. While a strictly instrumental picture of technique in the world may be efficient and effective, according to certain norms and values, it does not necessarily follow that everyone must see, understand, and read that world in the same way. Allowing this to be the case enables a deeper understanding of the challenge of engineering practice in all of its forms.

The reader may note that I have said little about "human error", malpractice or unethical behavior in my essay. As such, one might conclude that my analysis is seriously deficient and even evasive, i.e., writing off the search for a root cause as misguided suggests technical failure is never a matter of individual or corporate wrong-doing. The observation is correct but the consequent not: Of course engineers can be negligent, cheat, accept kick-backs from a supplier, alter the data, be oblivious to societal values, color the facts on the witness stand, etc. And so too corporate directors. My interest is in better understanding ordinary engineering practice not the pathological, the diseased, the abnormal. Most product and system failures spring not from ill-intention or evil doing but have their source in the mundane, everyday ethos of object world work. The unknown and under-determined nature of what we are about suffices nicely. Here is where I focus – on the collective enterprise of engineering design, seeking to explain technical failure when all participants work in accord with the norms and standards of the profession.

Another lacuna: Although the topic lies within my field of view, I do not address the contribution of law and legal processes to the definition of failure. In many cases the legal process is, in its quest to fix liability, a major ingredient in the definition and construction of failure. I acknowledge its importance but for me to do full justice to its role would require engaging a different world. That is a whole other chapter, if not an entire book. It suffices to say that one should not, in many cases, take the award of damages of the court as necessarily locating the root cause of failure.

What does interest me is the apparent disjunction between the world of law, of liability, of ethics, of moral judgement and the supposed value-free norms prevailing within object worlds. The way engineers keep values and machinery apart, made evident in many ways – the jokes they tell about lawyers, the use of the passive voice in all technical analysis, the puzzlement expressed when pressed to be socially responsible – resonates with, and is prerequisite to, their belief in the possibility of optimum design, of fault-free code, of unlimited technical progress and perfection. By looking at technical failure and allowing the possibility that it is oversimplifying, if not impossible, to reduce events down into a set of purely technical factors and another set of human, social, and/or legal factors we can better understand both why this chasm seems so deep and what might be done to remedy the fault.

If we do not pursue this possibility and shake off our naivete, we are left with a division of the world into two disconnected domains, (two cultures); the social on the one hand - where subjectivity, opinion, and values matter - and the technical/ scientific on the other - where objectivity, uniformity, scientific law and cold, value-free instrumental reasoning matter. Life within the former then depends on the needs, desires, and interests of peoples; life within the latter becomes banal, mundane, autonomous, purely instrumental - all object world work at all levels. This is not the world we live nor work in. We engineers ought to know better.

Notes.

1. If we accept that *knowledge is justified, true belief,* then I am claiming that engineers' beliefs in some instances – and important instances at that – are not true; they do not count as knowledge, even if justified in some way.

2. It is useful to distinguish between two kinds of technical productions: A *market place* or *consumer* product or system, one released, to the world at large, is one type; a *captive* product or system, whose use is under control of those responsible for its design - e.g., an industrial process - is a second type. The nature and possibility of conducting controlled experiments in the quest to explain malfunction differs significantly between the two. A third type of technological production can also be identified - the major construction project of a "one-of-a-kind" variety. Once in place, the freedom to change conditions in the pursuit of the cause of a failure of this third form is severely limited relative to the first two types.

3. It is often a legal process as well. This is a whole other world I hesitate to enter. While clearly what lawyers contend in court, what expert witnesses have to say, what jury's decide is the case - all of this is an essential piece of the social process. But it is a mistake to take a court definition of cause as definitive.

4. A fine report of the de-bugging process will be found in Kidder, J.T., *The Soul of a New Machine,* New York: Avon, 1981.

5. Public Information Office, Jet Propulsion Lab. California Institute Of Technology, Nasa, Pasadena, Calif. 91109. http://www.jpl.nasa.gov/releases/96/msolarpn.html (accessed 25 October, 2002).

6. This bit of the essay provokes questions about the relationship of model to artifact in the thinking of the engineer. In what sense is the model "equivalent" to the product? Is the product in "laboratory-like" conditions equivalent to the product "in the field"? How does a model construed as an "as if" picture of the system differ from one claimed to be a truer representation?

7. The reference for 'determined' here is the rigor of object-world analysis and prediction.

8. Bucciarelli, L. "Is Idiot Proof Safe Enough?" *International Journal of Applied Psychology* 2,4 Fall 1985.

9. Gettier, E.L., "Is Justified True Belief Knowledge?" Analysis, 23.6, June 1963 pp 121-123.

10. Dancy, J. *An Introduction to Contemporary Epistemology,* Oxford: Blackwell, 1985 p 25.

11. Bailey, D., "Comparison of Manufacturing Performance of Three Team Structures in Semiconductor Plants" *IEEE Transactions on Engineering Management,* 45, 1 February 1998.

12. ibid, p. 31.

13. Bailey, D., Personal correspondence 28 January 2002.

14. National Highway Traffic & Safety Administration, *Engineering Analysis Report & Initial Decision Regarding EA00-023, Firestone Wilderness AT Tires, Executive Summary,* iii, http://www.nhtsa.dot.gov/Firestone/firestonesummary.html (accessed 25 October, 2002).

15. Letter of Bridgewater/Firestone to NHTSA, 31 May, 2001.

16. NY Times, Tuesday, August 20, 2002.

4

Knowing that and how

In the last chapter, I explored what engineers don't know. This time I want to address what they *do* know. In particular, I am interested in the grounds for their beliefs and consequently their actions, their decisions, their designs. What is fundamental in engineering thought and practice?

First, a caveat: There are many different kinds of engineering practice; many different object-worlds; different disciplines for sure. Cutting another way, there are many different kinds of tasks engineers engage - design, diagnostics, research related to product development, manufacturing, project management, sales engineering, and let us not forget teaching - and engineers work in different industries, on projects of different scale - in budget, time, materials, market - and we might even allow for different national styles. I can not cover all of this ground so I will restrict my attention to work within an engineering object world, one I am most familiar with, and explore what counts as knowledge, what is fundamental there. I would hope that what I have to say is pertinent to a variety of such worlds, tasks, industries and cultures.

Another prefatory remark: I am a realist. I believe there is a material world apart from me (and you). But I also suspect that we can never know its true essences, "...the bare reality itself".[1] We see "...shadows on the wall of our cave", "now, through a glass darkly", but never the "thing in itself". We do fairly well, though, constructing general theories framed with mathematical rigor and working up phenomenological laws linking cause to effect - as well as thinking up cause and effect - and these suffice, at least for awhile, to explain the workings of 'bare reality'. They suffice in that they provide a set of coherent, socially valued and useful stories - explanations that enable us to make sense of the world around us in quite general terms and to remake the world to our liking in many particular ways.

History reveals how such theories and particular explanatory laws have been artfully conjectured, developed, derived, tested, and put to use. History also shows that alternatives are possible; our theories and explanations are never unique and, in one way or another, always a bit off. Whether we are progressing toward the truth is another question. This makes me a relativist but only to the extent that allows me to claim that people are different and see the world differently in significant ways, whether they be contemporaries or ages apart. Consequently, to understand the nature and status of scientific and/or engineering knowledge, one must study the social-historical context of its development and use.

Philosophers of science tend to ignore context; they appear to believe that knowledge is a matter of beliefs and their justification in terms of propositions and their logical consequences alone, uncluttered by social norms and values or cultural perspective. This serves them well in the rational reconstruction of scientific theories, in clarifying assumptions, in testing the coherent meaning of terms, and exploring completeness and the full reach of a theory but it says little about what it takes to *do* science, to build a theory, to shape an experiment to one's purpose.

In exploring what constitutes engineering knowledge from a philosophical perspective, any like attempt to uncouple knowledge, knowing, and know-how from its contexts of development and use is destined to be incomplete and unsatisfying. Two reasons for this: Contra science, engineering knowledge is not primarily textual, i.e., reported to the world in scholarly journals. While engineering faculty may publish their claims to new knowledge in the journals of their professional societies, outside the walls of the academy, within the firm, knowledge and know-how, which sustains participants in the design and development of new products and systems, are wielded in more varied ways: Internal memos, lab reports, parts lists, contracts and suppliers quotes have the form of traditional texts but to read these a knowledge of context is prerequisite. One can not assume, as one can in reading a physics abstract for example, that one's audience knows what one needs to know to grasp the meaning of such ephemeral productions. Drawings and sketches, bits and pieces of hardware, prototypes and suppliers' samples carry knowledge too; they are part of the language of engineering practice. This disparate collection of texts and things all enter into the reasoning about and thinking through of new designs.

Second, the kinds of "things" that enter into engineering discussions are a more varied lot than in science. True, scientific variables - well respected things such as force, displacement, temperature, time, charge, current, voltage, velocity, mass - all measurable and fit for rational explanation, play essential roles. But there are other variables, less well behaved and tamed - things like costs, margins of safety, legal codes and regulations, customer wants, aesthetics, ways of manufacturing, maintenance procedures - which enter into the accounts of engineers as they go about their business of knowing and of designing. While the scientist works within a single object world where the reduction of phenomenon to a well ordered set of variables of common measure is possible, indeed is the norm within any particular field, no such rendering and convenient abstraction is possible at the project level in engineering design and product development.

Because of this variety in content and form, to discuss engineering knowledge in the abstract, as embodied in traditional texts, as propositional and structured inference, is bound to, if not fail completely, at least neglect much that is important and significant in the reasoning of engineers. While the scientific-like theories engineers call upon display a logical coherence, the ties that bind ideas, concepts, and principles to the functioning product are not as determinate as one might expect or designers hope[2]. Only when joined with a sensitivity to the context of development and use, the philosophical study of the nature and status of engineering knowledge might prove both interesting and useful.

These are not problems of engineering alone but evidence of a more pervasive misconception about knowledge - that it is an entity in its own right, contained in books for all to acquire.

Knowledge

> *There ain't no such thing.*

A prevailing metaphor for "knowledge" suggests it is a material substance: We *gain* knowledge, *store it away* somewhere in our head; we *transfer* our knowledge to our students; some students claim that my course is "like *taking a drink* from a fire hydrant". In planning our courses, we decide what *material* we must cover, what to *leave out*, what to *keep in*. Knowledge is additive: the material of one course *builds* on another. We construct or *discover* new knowledge in our research and the value of our contribution is measured by, sometimes literally, the *number* of our publications. Knowledge can be *deep or superficial*. We know *more now than before*.

This way of speaking and thinking is mistaken: The metaphor - knowledge as "stuff" (solid or fluid or gaseous - solid is better than gaseous) leads us astray. I will continue to use the term, finding it hard to do otherwise, but I will try to speak of *knowing* rather than of *knowledge*, of an activity rather than of stuff, as this better fits my vision of its nature.

My critique is not original: Karl Popper describes as mistaken our "common-sense theory of knowledge" which sees our mind as "...a bucket which is originally empty, or more or less so, and into this bucket material enters through our senses... and accumulates and becomes digested"[3]. Popper would have us distinguish between two kinds of knowledge, subjective and objective. The latter "...consists of the logical content of our theories, conjectures, and guesses... Examples of objective knowledge are theories published in journals and books and stored in libraries; discussions of such theories; difficulties or problems pointed out in connection with such theories; and so on".[4] Popper is not alone in finding fault with the "container metaphor".[5]

I agree with Popper's critique in the main but find it useful to go one step further and make a distinction between information and knowledge. *Information* I take to be any representation, any human production which has been endowed by its authors with a disposition to provoke knowing. Thus, Popper's "theories published in journals and books and stored in libraries" constitute information, not knowledge in and of themselves alone. A drawing, a sketch, contains information. A prototypical fabrication of hardware is information - as well as machinery. A computer program, a differential equation, a list of specifications, a block diagram, the final product can be seen in some contexts as forms of information. A textbook, of course, contains information - but not knowledge.

Information *is* stuff. It can be conveyed, transferred from one to another, distributed, reproduced and done so accurately, without loosing a bit. Verbal expression, if fixed in a recording, is information; frozen in time, it can be distributed, replayed and interpreted (and mis-interpreted) just as other forms of information. You can add information to a drawing, subtract information by depressing the delete key. In certain contexts and for certain purposes one can even measure it as degree of orderliness.

What knowing is provoked by the particular information at hand depends only in part upon the author's intention. The author may intend to provoke certain knowledge, and the information may function to this end, but this is not assured. Nor can we rule out the possibility that knowing un-thought-of by the author might be the result. This is why I find Popper's concept of "objective knowledge" deficient; it implies that what is written in books and journals, none the less voiced and heard in a discussion, will be "read" in the same way by all. After all, that's what objective means. But while the author of texts, sketch, prototype, etc. would hope readings made would be in line with their purpose, this may not prove the case. We can ask then, how does the information function? Is it in accord with the author's intent? At the same time, we should allow for a creative reading that may be in line with the intention of the author but extends or embellishes. The author may even confess that he or she (or they) had never thought of that. Creativity springs from around the edges of words.

In sum, the same body or stream of information can provoke different readings, different knowing, by different persons. (Object worlds again). The meaning and significance of what I say may not be the same as what the person next to you derives from my words. What one student claims to know after reading from my textbook will not be the same as what another student learns from the same selection, though we pretend they all know the same. We go to the movies; you see the film event as fore-shadowing and know what is going to happen next; I miss this entirely - though I was not yet asleep. The instructions for assembling the do-it-yourself product, e.g., a backyard barbecue, may be interpreted one way by one person, another way by his brother[6]; both, if successful, know how to do it. Wittgenstein asks me to continue the series, to extrapolate on the basis of the information contained in the sequence of numbers, and I go one way, he goes another[7]. And Clifford Gertz has convinced me that I must be careful when I wink in a foreign land[8].

While intention is essential to the production of information and the provocation of knowing, the meaning construed may not be as intended, no matter how much effort the author puts into his production. Of course, this is not always the case. Because so much of common discourse is conducted in terms of standard, culturally saturated, forms and expressions, the author need do little work giving form to intention as information in many standard settings and situations. Searle's analysis of speech acts is all about common forms.[9] But when the situation is new, the intention not standard, the ground not covered - as in the design of the new - then the task of transforming intention into information which provokes the knowing one intends becomes real, formidable and non-trivial. Authoring remains no light task.

This distinction of information from knowledge also makes knowledge, as knowing, an event in time requiring action on the part of those who provoke and those who would come to know. The same information - e.g. text, artifact, drawing, photograph, signal - in two different temporal settings can provoke different thinking and knowing. Newton's axioms provoked certain interpretations in the wake of their appearance. They neatly provided a way to calculate and understand the motion of seemingly all heavenly objects. Today they provide the basis for knowing about all kinds of terrestial as well as celestial phenomena. The concept of force in the 17th century was understood and used differently than today. So too the idea of "mass"[10]. To claim that Newton and his contemporaries "knew" these concepts the same way as scientists and engineers do today is mistaken[11]. Again, this is not to claim that we can not, do not, have a framework for talking about the differences in understanding displayed in different historical periods. Our grand worlds are not incommensurable. Still one might claim that Newton worked within a different object world - of force, position, velocity, impact, momentum - than applied mechanicians do today. And while we can always construct some anachronistic comparison from today's perspective, at the same time we ought to confess how difficult it is to see things otherwise: One can wonder how truthful to the author's intent any historical explanation can be or, what is much the same thing, whether we can truly re-live the historical essence of Newton's Newtonian Mechanics though sufficient information is there for all to see and read. One need not juxtapose different historical periods in making the point: Two scientists who are contemporaries may have quite different interpretations of the meaning of a proposition.

De-coupling information from knowledge, together with the idea that different persons may know differently after engaging the same information, also enables me to accommodate those who claim that artifacts convey or contain knowledge. "Thing knowledge" [12]from this perspective is then what one reads out of the materials at hand. But again, there can be different, superficial, deep, or even erroneous interpretations by different persons.

Finally, it naturally follows that any attempt to distinguish engineering knowledge from scientific knowledge must move beyond the comparison of texts and other forms of information. Only through consideration of what engineers *do* and what scientists *do* can one come to distinguish what engineers *know* from what scientists *know*. Much of engineering information has the same appearance, is the same in this respect, as information available to and relied upon by scientists. But what engineers know and what scientists know is not revealed there alone. While we can speak of engineers applying scientific knowledge, it's best to get our mind off the stuff and take a look at how and why and when it is applied (but not like paint).

Structural Engineering Object World

Engineers come in different kinds; they address different tasks, of different scale and complexity, have different responsibilities, competencies and interests.

Boundaries between kinds can blur: Some engineers *do* work like scientists; others like managers. Some interact daily and intimately with hardware, others only deal with objects as fantasies, in software. But all who have an accredited engineering degree do share one important thing in common - a science-based, university education.[13] This frames their thinking, their knowing, their doing whatever their position. In this section I focus on object-world thought and practice with a particular discipline, one common to university programs in mechanical, civil, and aerospace engineering - a particular discipline I am most familiar with, namely structural mechanics.

Physical, material structures come in different kinds. There are *truss structures* (we saw one in a previous chapter); there are *frame structures* made out of beam elements; there are *cable structures*; *plate structures* and *shell structures*. All these various forms, when subject to external loads, experience internal forces and stresses; and they deflect and deform. They may vibrate and resonate, quake at specific frequencies. They also can fracture, flow like a plastic, corrode, crack and fail due to cyclic loading - what's known as fatigue. They can absorb energy - think of the crash worthiness requirements in the design of automobiles. They expand when heated and this can engender excessive internal stresses if the structure is over-constrained.

In designing a structure within a particular context, at least some of these phenomenon must be addressed and an explanation of behavior constructed: If I apply an end load of a given magnitude to the end of this cantilever beam, will it fracture? Where will it fracture? How much will the end deflect if it does not fail? If I suspect my design might fail due to fatigue, what tests should I set-up and run to verify my design? What thickness should I specify for the walls of a soda can to ensure its integrity both under internal pressure and the firm grasp of a hand when opened?

There is a fundamental theory for answering these and similar questions: The mathematical theory of elasticity. Perhaps I should say "theoretical framework" rather than "mathematical theory". There *is* a mathematical theory of elasticity, but this suggests that all questions are addressed and all problems solved by rigorous deductive-nomological derivation of specific laws applicable to the particular phenomenon at hand from this theory. While, in principle, this may be possible, this way of envisioning matters would be a gross violation of both the historical origins of particular theories and of how rules and instrumental relationships are re-construed in practice.

The full theory of elasticity is taught in schools of engineering, usually at the upper-class and graduate level but undergraduates see the same concepts and principles introduced in their study of particular structural elements and forms, e.g., in the analysis of the behavior of trusses, cables and beams and frames. The theory is laid-out in certain canonical texts[14]. Internal histories have been written[15]. The giants upon whose shoulders we stand include not just Newton but Galileo, the Bernoullis, Leonhard Euler, Lagrange, Laplace, Poisson, Navier, Cauchy, Young, Stokes, and others. This is hard-core, mathematical-scientific stuff.

Now while instruction in the schools of engineering does not do justice to the full reach of the theory, and hardly says a word about its historical development, all courses in statics and strength of materials, in the behavior of solids and structures, in engineering mechanics for structures, as well as the theory of elasticity itself, speak in terms of the same theoretical objects - force, moment, stress, strain, displacement - and refer everything back to three fundamental requirements; the equilibrium of forces and moments, compatibility of deformation, and constitutive laws which relate force to deformation or stress to strain.

In what follows I intend to illustrate the way engineers think and work within this theoretical framework and put concepts and principles to use in their teaching, in predicting or diagnosing failure, and in justifying their designs. My purpose is to explicate the fundamentals of engineers' ways of knowing, to explore how theoretical objects, laws, and claims mix with mathematical representation and method and with the observed, sometimes measured, behavior of real structures in the thinking and doing of engineering. I consider an excerpt from an engineering textbook; a historical tract of importance; and the design and development, building and use of an instrument. I pay particular attention to the language engineers use and the narratives they construct - stories intended to provoke knowing, not just by others, but in a reflective mode, by themselves as well, working alone. I am interested in exploring how the narrative, within which the "science" is embedded, does its work. Through these examples we begin to see what is fundamental in their knowledge claims and how that justifies their designs. We discover too how information can be interpreted in different ways, extended, codified and archived.

An Engineering Textbook

My first example is drawn from a well respected textbook on our subject[16]. It concerns *the derivation of the necessary conditions for static equilibrium of a solid body, viewed as an isolated system of particles.* It's purpose is to relate the fundamental principles that necessarily must hold for an isolated body to be in static equilibrium, back to a still more fundamental picture - that of the body as a collection of particles to each of which we apply Newton's laws.

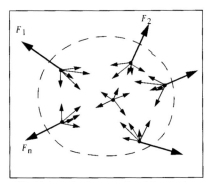

Fig. 4.1 An isolated system of particles showing external and internal forces.

Already my language becomes opaque, coded. The information contained in an engineering textbook has this mystifying quality: The words look like English - mathematical expressions notwithstanding - and the figures look simple enough, but something strange is going on here. *Equilibrium, particles, body, Newton* - these are all familiar terms. But the way they are used, their collective impression is of another world - an abstract, timeless, neat and proper world where static equilibrium is not only possible but necessary. A world of too simple figures - minimalist in content - all described in a passive voice. There are no human agents here, only gravity, continuity, equilibrium and a garden of a variety of materials. In what follows, I italicize passages of textbook type so the reader will know when we have entered and exited this world.

Our story begins with the picture above[17]. *The dashed line stands for the boundary of our isolated system of particles. All interactions with the world outside of the body are replaced by force vectors (or moment vectors). The force vectors acting on the body due to this isolation are indicated.* From here on, we can forget about the rest of the world[18].

Inside the boundary we see six particles. These entities are featureless in the sense that they have no color, no particular shape, nor even size, although they must be sufficiently small. At one time it was thought that their shape determined how they interact, and so their collective behavior[19]. But our narrative has no need of shape, color, nor size - just "smallness"; but even the measure of the latter can remain unstated. What the particles *do* have is the property of exerting a force on their neighboring, like particles. They are *centers of force*.

Of course there are more than six particles in a solid body; generally we think of *a continuous body as having an infinity of particles. Each of these is subject to external forces - e.g., gravity. Some will experience other externally applied forces; all will be subject to internal forces due to the other particles. These latter, in accord with Newton, occur in equal and opposite pairs.*

This neglect of the infinity of particles is one form of "all other things being equal": All the other particles are taken to behave just as these six behave. Indeed, there is nothing sacred about six; the authors could have shown three, or four, or more than six. But there is another kind of "all other things being equal" going on here in the sense that whatever else might be described as properties of solid bodies or their constituents, beyond this sparse description in terms of small particles and the forces acting between them, is irrelevant to establishing the necessary conditions for the equilibrium[20]. Of course, what is not relevant encompasses an infinity of properties and things. To describe all that does not matter in the rest of the world is not feasible The textbook narrative presumes the reader already understands the irrelevance of most of the world; that students already have a sense of the furniture of this particular object world in engineering mechanics from their studies in physics and mathematics; they must already understand what kinds of entities are allowed into this world; they must already speak the rudiments of this language.

Now for static equilibrium, again in accord with Newton, the sum of all the forces acting on any particle must vanish so the sum of all the forces shown in the figure must vanish. But since all of the internal forces occur in equal and opposite pairs, they sum to zero. So we are left with the result that the vector sum of the external forces must be zero

$$F_1 + F_2 + \ldots\ldots + F_n = \sum_j F_j = 0$$

There is something magical about this process; by introducing and adding zero sum pairs of internal forces we appear to derive the first requirement for static equilibrium which the external forces must satisfy. How can introducing *nothing*, give us something? Alternatively, why go to all this trouble? Why not simply posit the requirement - *for equilibrium, the vector sum of the external forces must be zero* - as fundamental and be done with it? One respected critic has argued this way.[21]

Continuing, *considering the total moment of all the forces about an arbitrary point yields the result (again because the internal forces self destruct) that the total moment of all the external forces about an arbitrary point must be zero.*

$$r_1 \times F_1 + r_2 \times F_2 + ... + r_n \times F_n = \sum_j r_j \times F_j = 0$$

Here then are *the two requirements for static equilibrium of a rigid body.*

The simplicity of these equilibrium equations is to be noted - simplicity in the sense that there are but two species of entities that enter into them - force and position. This simplicity corresponds to the poverty of our picture showing the particles constituent of the solid body. These mathematical relations go hand-in-hand with the picture and with the narrative to, all together, constitute the necessary conditions for the static equilibrium of a rigid body. The equations alone do not suffice - too ambiguous; what are we talking about? The picture alone does not suffice - too sparse. Newton's laws alone do not suffice -too general. But all of these bits, together with talk about boundaries, external forces, reference positions, and presumptions about the rest of the world, make a coherent and useful narrative.

What kind of knowledge is this? Abstract, certainly; universal too. Its use of the passive voice and dismissal of human agency assures it's communal acceptance. It is intended as dogma, as fundamental concepts and principles, that will empower those who learn the language and enter this world with the right perspective and with the competence and confidence to analyze, to diagnose, to design any structure conceivable in its terms. The sense is that if the student learns but these *two requirements for the equilibrium of a solid body*, then he or she possesses what is sufficient as well as necessary for the solution of any problem he or she might confront in the engineering mechanics of structures[22].

While not all textbooks offer this "derivation" of the two requirements for static equilibrium, the intentions of other authors is the same: to reveal the power and general applicability of such a sparse set of concepts and laws. It is reduction at its finest, yielding an efficient and all encompassing basis for analysis within the world of mechanics. There is clearly a principle of economy of thought evidenced here: We need but faceless particles together with but one fundamental law to explain all; *equilibrium of a solid body is but a consequence of equilibrium of forces acting on a particle.*

Textbooks are written for teaching purposes, of course. Some have an orientation toward the "applied", others more toward "theory". Textbook styles change with the times, though theory stands a good chance of holding a steady presence. Few, if any, reflect on the historical or philosophical foundations of what they present as fundamental, and with good reason: For engineering knowledge is aimed at doing, at making, at producing the new; its focus is not the past - to which we now turn.

A Historical Development in Theory

Not all textbooks include this derivation from first principles. In fact its legitimacy can be called into question on several grounds. I have already referred to one critic's complaint; I will focus on another related way in which it is a bit off. It's not that these two equilibrium requirements are not "true"; every student learns quickly that they must be accepted as fundamental and applicable if they are to solve any of the assigned problems in engineering mechanics. It's that, taking this vision of interacting particles as a theoretical representation of the behavior of a solid when it is allowed to deform, can lead you astray. The picture does not work for a collection of particles that is subject to external forces, subsequently deforms, and one seeks to determine the displacements of all the particles in the deformed state.

Navier tried this in 1821. He modeled a deformable, elastic solid in accord with the picture. He looked at an arbitrary particle - what he called a molecule - and considered the equilibrium of that particle due to the forces of all the other surrounding particles and the external force it was subject to. So far everything is in accord with our textbook figure[23].

He then allowed the particles to move relative to one another - the body to deform. He posited that the force acting on any isolated particle due to another particle was linearly related to the change in distance between the two. (Navier is showing good engineering sensibilities here. Try a linear relationship first, see what that gives you). He did not know what this force law was but posited that its magnitude depended upon the original distance between the particles as well as the change in distance between them. In fact, following Laplace, he claimed that it was only *"sensible at insensible distances"*. This enabled him to forget the boundaries of the solid and replace his summations by infinite integrals which he claimed gave a finite result. All that remained of this unknown force law in his final system of equations for the displacement components of the molecule - here designated by u, v, and w - was a constant, λ. In modern notation, his results take the form:

$$2\lambda \cdot \frac{\partial}{\partial x}\left(\frac{\partial u}{\partial x} + \frac{\partial v}{\partial y} + \frac{\partial w}{\partial z}\right) + \lambda \cdot \left(\frac{\partial^2 u}{\partial x^2} + \frac{\partial^2 u}{\partial y^2} + \frac{\partial^2 u}{\partial z^2}\right) + F_x = 0$$

These equations look very much like the equations of equilibrium expressed in terms of displacement for a linear, elastic, homogeneous, isotropic body appearing in textbooks on the Theory of Elasticity. There is only one thing wrong and that is fundamental: There is but one constant, λ, appearing in these equations. Today's theory requires two, independent elastic constants. The equations now have the form[24]:

$$(\lambda + \mu) \cdot \frac{\partial}{\partial x}\left(\frac{\partial u}{\partial x} + \frac{\partial v}{\partial y} + \frac{\partial w}{\partial z}\right) + \mu \cdot \left(\frac{\partial^2 u}{\partial x^2} + \frac{\partial^2 u}{\partial y^2} + \frac{\partial^2 u}{\partial z^2}\right) + F_x = 0$$

Navier's work did not avoid critique. A contemporary found fault in the way he transformed his summations over discrete forces into integrals over the continuum. But addressing this complaint would not introduce another constant (nor was it a complaint with merit). One might think that the deficiency in Navier's theory would have been discovered through experiment. Yet this proved quite improbable if not impossible. For Navier's relationship can be obtained from the correct set by setting λ and μ, the two constants, equal, - an equivalence approximately true for many common structural materials. While these "correct" equations of force equilibrium in terms of displacement were established not too long after Navier presented his memoir to the Institute, controversy over the number of independent, elastic constants persisted over much of the 19th century.

I am not going to try to rationally reconstruct how and why Navier went wrong, what that might mean, or how the *correct* theory subsequently evolved[25]. What is worth comment is the relationship of the mathematical, analytical content to the basic physical principle he employed - that of *sensible forces at insensible distances*. The mathematical content is sophisticated, state of the art of the time; the physical principle seems too simple, too cryptic to lead anywhere. But there was much there.

Laplace was the respected author of this notion, a principle he and his proteges usefully employed in the analysis of a wide range of phenomena. The "correct" provocation or stimulus would have the reader understand that *one need not know the explicit dependence of the force acting between any two particles upon the distance between the particles in order to construct a useful analysis*. Again, magic: We seem to posit "nothing" in the sense that you don't need to get wrapped up in details about the nature of this internal force, its source, how it varied with distance, size or shape of the molecule, etc., in order to derive a coherent, logically consistent and useful explanation of the deformation of an elastic solid subject to externally applied forces[26].

Today's textbook development of the theory of elasticity, the correct theory for deformable solids, offers the same generality but presents a quite different picture of a solid body. This theory, due primarily to Navier's contemporary Cauchy, sees an elastic solid as a homogenous continuum. No molecules here - nor particles of any shape or size whatever. Internal forces are construed, not as forces acting between particles, but as stress, like pressure, as a force per unit area. For example, a tensile stress at a point within a body is defined as the limit of ratio of the force acting perpendicular to a differential element of area to the area as that area becomes vanishingly small. You might think that if one focuses on the vanishingly small one might encounter an atom or two and begin to wonder what happened to our material world - of molecules, or atoms, or grain boundaries and slip planes. No matter; we engineers accept this particle-less picture and work with it to our advantage. It is not only essential to our object-world analyses in the design of structures but, while scientists may consider the theory of elasticity a matter of the past, new extensions and interpretations, often motivated by the behavior of new materials in new configurations and mixtures prompt research within its domain. While the urge to speak in terms of universals, general concepts and a minimum of principles is real among engineers, this reflects, not a thirst for scientific truth, but the need for a sufficiently rich and expressive (and efficient) language to carry on with the job of predicting the behavior of real structures.

Here is a significant distinction between the status of theoretical knowing among scientists and among engineers: Scientists are concerned with the explanation of nature's phenomena (what's left of it). Engineers are concerned with understanding the built, the engineered world. This entails that engineers can take a certain detached view of the epistemological status of the theories they think, design and, build with. Their stance is pragmatic. Verification is a matter of how well their productions work in accord with their predictions. So if they can deduce useful relationships about the behavior of structures from a an abstract mathematical representation of structural elements as continua, that will suffice. They are perfectly happy with statements like: A beam behaves *as if* it were a continuum - made up of molecules, atoms, fundamental particles; they care not. All of this is context dependent.

A critical reader of my juxtaposition of these two different representations of the workings of a solid body might object: Our textbook authors are explicit in stating their model is meant to be of a rigid body - one in which the distances from any particle to any other does not change when subject to externally applied forces so I am not justified in applying it to bodies that do not respect this condition. The point is well made, but overly restrictive. For if we insisted that this criterion apply to the reading and extension scholars make of the productions of their predecessors, there would be little if any, even normal progress in the sciences, none the less in engineering. A more legitimate complaint would fault my association of these two events drawn from two quite distinct contexts - a 20th century American textbook and a memoire of the French Academy of Sciences, dating from the early 19th century. The former objection accepts the apparent timelessness of engineering and scientific knowledge: Context, from this perspective is irrelevant. The latter objection accords context its due and is worth further consideration - but we leave that for another time. I now turn to a still earlier historical period and consider the construction of knowledge about the behavior of a particular structural element - the cantilever beam.

Galileo and the Fracture of a Cantilever Beam

Scientists and engineers both claim Galileo as ancestor. He is best know for his defense of Copernicanism. I am concerned here with his more down to earth studies - that which we find in the Second Day of his "Dialogues Concerning Two New Sciences" where he addresses some truly practical questions about the breaking of beams.

The figure shows his cantilever beam. He wants to know how big the weight at the end can be before the beam breaks. (Don't be distracted by the wall. You are to assume the wall will not fail, in fact remains rigid. *ceretus paribus*).

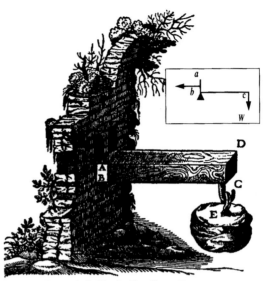

Fig. 4.2 Galileo's Cantilever Beam

In engineering parlance, he "models" the beam as a lever, an angular lever. I have added a figure, corrupted his drawing, showing the way I read the information provided in the English translation by Henry Crew and Alfonso de Salvio of his *Dialogues Concerning Two New Sciences*.

The lever is an idealization, abstract in more ways than one. The lever is a rigid body; it doesn't deform; the beam remains straight. It is assumed weightless, at least at the start. The weight at the end is taken to be concentrated, acting at a point C. I have shown the "....resistance to fracture which exists in the thickness of the prism, i.e., in the attachment of the base BA to its contiguous parts..." as an arrow directed to the left (the contiguous parts pull back on the end of the prism) and labeled it R. I have shown a fulcrum, about which the lever can rotate, at b.

For equilibrium of the lever, the ratio of R to W must be the same as the ratio of the length bc to one-half the length ab. That is[27]

$$\frac{W}{R} = \frac{1/2 \cdot ab}{bc}$$

So if you know what force is required to break the prism by pulling it along its axis, putting it in tension, namely the "absolute resistance" here designated by R, then this relationship tells you what end-load the same prism, working as a cantilever, can sustain. This is Galileo's main result translated into today's proper language of engineering mechanics applied to structures.

Galileo then considers how this result changes if the weight of the prism itself is to be taken into account - in which case half of the prism's weight is to be added to the end weight, W, or, we can say that the endload to produce failure will be reduced by half the weight of the prism. (The prism's weight, assumed uniformly distributed along its length, effectively acts at half the distance bc from the fulcrum).

He goes on to consider two prisms of different cross-sectional area and claims "....no one can doubt" that the strength, or absolute resistance is proportional to the area "...because...the number of fibres binding the parts of the solid together in one cylinder exceeds that in the other cylinder" as the ratio of the areas. With this he develops a multitude of particular laws which account for differences in failure load due to differences in each and every dimension of the beam.

This excerpt serves as an excellent illustration of the use of abstraction in the idealization of real structures. Galileo "sees" the beam as a lever; it's like he has put on a special pair of eye glasses which enable him to see beyond or through what any common person would see, i.e., the beam as it is pictured in his text, the beam which most philosophers would see and credit to external stimuli.

Galileo also sees a uniform distribution of fibers over the cross-section. In this lies an implicit recognition that the critical theoretical concept, one essential to predicting when the beam will fail, is the force per unit area - what Cauchy would later call stress. If this force per unit area exceeds a certain value, fracture of the beam at its root will ensue.

His dialogue is very much engineering in tone and topic. It includes:

- A mathematical relationship derived within the context of the overarching framework of theoretical objects and concepts - force, moment arm, moment, and the requirements for equilibrium of a rigid body.

- This embedded in a narrative which lays out what is to be taken as cause (the end weight) what as effect (the fracture of the beam at the root) and how the cause is related to the effect through the principle of the lever and another narrative about how the failure load in tension is proportional to the cross sectional area.

- A statement about failure condition: If you can determine a material's absolute resistance to breaking in what we would call now a tension test, then you can predict when cantilever beam of that same material will fail.

- An exploration of the breadth of application: The analysis and model applies to beams of all kinds "...glass, steel, wood or other breakable material".

- The devolution of a set of consequences for circular cylinders as well as rectangular prisms of different dimensions. How long might the beam be before it fails of its own weight? What if the beam is supported at both ends rather than at one end alone: He even argues why giants could not exist if simply seen as scaled up versions of ordinary persons. In engineering we strive to develop correct scaling laws.

Despite the insight he displays, his result is incorrect when judged against today's theory of the elastic behavior of solids and structures, and engineering beam theory in particular. While his results are dimensionally correct the factor of

1/2 is wrong. This ought not to detract from the importance of his historical "contribution" nor negate the significance of his insight - that the failure of the beam is due to what we now call an internal bending moment - all else being equal, e.g, the fastening of the beam to the wall at its root is secure.   His analysis remains correct and useful in scaling up a beam whose fracture load you already know. More to our purposes here, Galileo's strategies for coming to know when a beam might fracture are akin to those adopted by today's engineers concerned with the behavior of structures.

Measuring lift to drag ratio.

My next, and final example is more modern. It concerns apparatus to measure the forces acting on an airfoil, mounted in a wind tunnel. Our question is again, what is fundamental, what must one know and know-how in order to understand why and how this functions - so that one might design, build and successfully employ the device.

Fig 4.3 A wind tunnel "drift" balance[28]

In the photograph, the vertical black bar is in fact the airfoil whose effectiveness we wish to measure. It is rigidly attached to a horizontal member, which in turn is attached at its ends to two brackets. The brackets are free to rotate about vertical axes. These axes are supported by the two posts of greater diameter. Another member connects the two brackets at their ends away from the test specimen.

A schematic drawing shows better how the members are connected and the principle of operation. Here is the schematic, taken from a NASA web site[29].

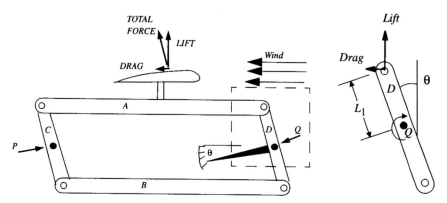

Fig 4.4 The wind tunnel "drift" balance - Top view.

The view is from above, looking down onto the mechanism. At the top of the figure we see the cross-section of the airfoil test specimen. (The airfoil extends in and out of the plane of the paper). To the right is shown an enlarged view of the member D and includes a picture of the (internal) forces acting at the point where D is pinned to the member A.

Air flows at some velocity from right to left. As it flows over and around the airfoil, the specimen experiences a force which can be decomposed into two com-ponents - a Lift force perpendicular to the airstream and a Drag force in the direc-tion of the airstream.

The test specimen is rigidly attached to the top member A of the mechanism. The four circles at the corners are pins, made as frictionless as possible so that the parallel bar motion is not restrained in any way. The dark circles at P and Q are the thin cylindrical, supporting shafts, the axes about which the two brackets, labeled C and D here, are free to rotate. These, in turn, are fixed to a base at the floor of the wind tunnel. A pointer is fixed to member D - it always remains perpendicular to member D.

In the presence of an airflow of a particular velocity the mechanism will rotate counterclockwise as shown until equilibrium is established. The device will mea-sure directly the ratio of the Drag component to the Lift component. In fact, it can be shown that the ratio of the Lift force to Drag force is a relatively simple func-tion of the angle, θ, namely

$$\frac{\text{Drag}}{\text{Lift}} = \tan\theta$$

We can deduce this relationship from the requirements of static equilibrium applied to the system, here modeled not as a continuum but as a system composed of four rigid links. In contrast to Galileo's analysis of the cantilever beam where I drew one idealization, a number of idealizations, one for each of the four members of the mechanism must now be constructed. For example, an idealization of the top member *A* yields:

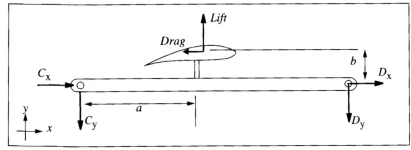

Fig 4.5 Idealization of top member A

In this, we presume that the pins at *C* and *D,* as frictionless, can not transmit any torque. Our task is to determine the "internal force components" C_x, C_y, D_x, D_y, in terms of the Lift and Drag forces and the distance measures, *a* and *b*. Equilibrium requirements for this member yields three equations relating these unknown components to the Lift and Drag force components. But since there are four "unknown" components, more is required: The more is supplied by making idealizations of the other three members.

As this is not meant to be a mechanics textbook, I spare you the details, and simply announce that this accounting procedure eventually provides a sufficient number of equations to yield the sought after result, namely that the tangent of the angle theta is equal to the ratio of the Drag force to the Lift force. Curiously, we find though that one can not obtain a sufficient number of equations to solve for internal force acting in member B. Yet the relationship between the lift to drag ratio and the tangent of the angle does "fall out" of the analysis.

In this example, as in those before, we can explicate knowledge, engineering knowledge, with respect to two different contexts: One, the object-world of contemporary engineering mechanics (and aerodynamics). Two, the historical, more worldly context of the development and first use of the device.

Our report so far has focused on what Ryle would term *knowledge as knowing that*[30]; knowing as abstract ideas, timeless concepts and relationships expressed in mathematical, symbolic form and as analysis of function embedded in a narrative where the passive voice prevails; e.g., "In fact, it can be shown that..." "...We know that... if the member *AB* is to be in equilibrium - not move - then the resultant force and resultant torque on the member, isolated from the rest of the world must vanish - this in accord with the two equations "derived" from Newton's laws as in our textbook...." We know too that the equilibrium requirements are necessary and may be sufficient to justify the proposition: the ratio of the drag force to the lift force is given by the tangent of the angle theta. That they may be insufficient must be allowed as the structure may be overconstrained.

There is little in our report about knowing in the sense of Ryle's *knowing how* - as in practice and doing and in the manipulation and shaping of hardware and constructing explanations of a more worldly sort about function. While the context for *knowing that* stands apart from the ordinary world - it's a place of frictionless pins, of undisturbed, uni-directional airflow, weightless supporting members, and truly straight lines, truly dimensionless points and appears timeless, out of time, out of history - the context of *knowing how* is a different sort of place. In this case it's the world of the Wright Brothers and the first decade of the 20th century. Here the world appears more familiar - an ordinary world of bicycle spokes and hacksaw blades, of imperfect welds and pulsating air currents, and hopefully negligible deviations from straight and true.

We turn now to the historical context of development and first use of the instrument.[31] Within this context we would like to know how the Wrights did what they did - engineer an ingenious device for measuring the ratio of the forces of lift and drag. How did they proceed? What did they know how to do? What must they have known? How did they imagine the device? What was source of their ideas? What justified their beliefs and actions; their trials, their designs?

Here now we move outside the confines of our neat and tidy object-world of up-to-date engineering mechanics. In fact, if we truly wish to recreate the past, it's best if we unlearn some engineering mechanics - if that is at all possible. We must strive to distance ourselves from ourselves, i.e., from the way we read and understand the device. For what we seek to understand is how the authors' imagined and constructed the instrument and in our quest we should not presume that they spoke the modern language of engineering mechanics as we do today, or at least not in the same way we do today. The challenge of writing justified and true histories is real; historians worry and write about what's required.

Joseph Agassi, for example, set two conditions on how the history of science, and by implication, the history of technology, should be written.

> The first maxim of enlightened or broad-minded historiography should be this: any interesting or stimulating story is good, and should count as history if it fulfils two conditions: (a) it does not often violate factual information easily accessible to its author, and (b) it does not present historical conjectures as if they were pieces of factual evidence[32].

These strictures are meant to provoke more than enlighten. Agassi is saying that good history is not just a record of historical facts but a good story. Even apparent and easily accessible facts can be violated, ignored, twisted around if the historian judges the case otherwise. Conjecture is essential too, but be on guard: One must not mix historical fact with the historian's constructions. Even so, Agassi's rules grant the historian considerable freedom to play with the facts in their reconstructions.

While Agassi's "positive views" - he calls them that - lay out what the historian should *not* do, Collingwood is more explicit in describing what is needed in order to make an interesting and stimulating story: Interpolation, as well as critical assessment of the facts and source materials, is necessary. One must use one's imagination in fleshing out the past. But the imagination is to be employed in ways "...not ornamental but structural. Without it the historian would have no narrative to adorn."

The imagination, that 'blind but indispensable faculty' without which, as Kant has shown, we could never perceive the world around us, is indispensable in the same way to history: it is this which, operating not capriciously as fancy but in its a priori form, does the entire work of historical construction.[33]

The sense one has here is that of necessity, not of possibilities. Just as through induction, the scientists develops a coherent theory that can be put to the test, the historian must employ his inductive powers imaginatively to fashion his story then confirm his construct through additional facts and others' stories. Imaginative re-enactment of the past and the past thinking of history's agents is Collingwood's way of doing history. The proper task of the historian is to penetrate "... to the thought of the agents whose acts they are studying."

In all of this, both Agassi and Collingwood would agree that our reconstructions of past thought and deed, should be in accord with the standards, beliefs and norms of the historical period we study. There is no contradiction here, only the challenge of keeping historian's conjecture apart from historical fact. A historian's responsibility is to re-enact using only the props and stagings of the past. We must in this continually struggle, in Agassi's words, to avoid "being wise after the event". He points to Koestler's recommendation that when approaching the past "...we see ourselves as children" - which Agassi, however, sees as insufficient.

Our task then is reconstruction: We seek to lay out a coherent, rational sequence of ideas and relationships in a story that explains how the Wright Brothers did what they did. We rely, in this, on a coherent, rational theoretical model of how the instrument must, should, would work. The latter form of representation can be made watertight; but the former not. That is, the rigor of object-world theory should not be taken to justify full belief in the historical narrative I construct. Just as the rational, logical, efficient workings of a finished product can lead one astray in attempting to reconstruct the design process, so too, an equilibrium analysis of this historical artifact should not be equated to how it came to be.

We attend to how the Wright Brothers proceeded, focusing on the apparatus they came to know and use so effectively, drawing upon original source materials, fortunately published in book form.[34] Our interest is in the knowledge they employed in their fashioning and reasoning about the "drift" balance. While my subject is object-world knowledge, my telling is not in the passive voice, disinterested, value-free, written around some mathematical, symbolic expressions which evaluate to the same result wherever one might be in space or time, but rather a narrative true to the context of the times which attempts to say how they did what they did. However, although the Wright Brother's accomplishment is worthy of much praise, I will not say much about the weather (e.g., "On a wind swept dune, one bright morning in Kitty Hawk..." etc.) There are limits to what is socially relevant to object world work. My history is intended to be a history of ideas, of engineering, object-world knowledge and knowing.

I start with a letter from Wilbur Wright to Octave Chanute dated September 26, 1901 in which the author describes an experiment designed to verify the data, reported by Otto Lilienthal, a German predecessor who attempted flight (and who died in a crash of one of his gliders), describing the lift that one might obtain from an appropriately curved and inclined wing surface.

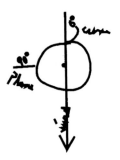

In Wilbur's letter, we find a simple figure, a hand sketch, included as an essential part of his communication:

> I am arranging to make a positive test of the correctness of the Lileinthal coefficients at from 4° -7° in the following manner. I will mount a Lillienthal curve of 1 sq. ft. and a flat plane of .66 sq. ft. on a bicycle wheel in the position shown. The view is from above. The distance from the centers of pressure to center of wheel will be the same for both curve and plane. According to Lilienthal tables the 1 sq. ft curve at 5° will just about balance the 66 sq. ft. plane at 90°. If I find that it really does so no question will remain in my mind that these tables are correct. If the curve fails to balance the plane I will cut down the size of the plane till they do balance. I hope to make the test on the first suitable day[35].

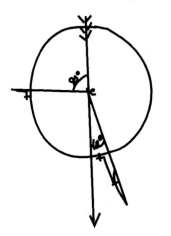

A week later, Wilbur reports the results, again in a letter to Chanute. Another new figure accompanies his report.

They found that Lilienthal's predicted lift would not balance the plane. To figure out why this was the case, they experimented with the airfoil's orientation until they did achieve a balance of the two. (Though they said they would cut down on the area of the plane, it evidently was easier to alter the angle of incidence of the curved surface). They found that they could balance the flat plate but only by increasing the angle of attack (to 18°).

If we stop here and ask what the Wright's need to know, what they knew (both that and how), they clearly know the concept of equilibrium of torque and how to take advantage of it. Like Galileo, they are balancing an angular lever. This is the underlying form, the object-world principle they apply (but not like paint). And it is this concept which threads through all of their subsequent measurement work and the design of their instruments.

They knew as well how to take advantage of what was construed as an exemplar, a common, established case - that of the force on a flat plate set perpendicular to the direction of the wind. Their measurements, up until the last instance using the drift balance, were all made relative to the force acting on the plate. Soon enough, however, they were to question what was known about the *absolute* value of the force on the flat plate placed in an airstream, knowledge embodied in a constant labeled *Smeaton's coefficient.*

At first, they had tried setting out their bicycle wheel in some fixed position relative to ground in a "natural wind" but this proved unworkable.[36] This did not give satisfactory results. In the second letter they report how they had recourse to mounting their apparatus over the front wheel of a bicycle to assure themselves of a more regular and sustained air flow; With this new mobile arrangement, they rode off "...from 12 miles per hour" to ensure a steady airflow - as steady as possible at any rate. To null out the effect of any natural, arbitrarily directed wind, they purposely rode

> ...at right angles to the wind so that the natural wind was first on one side and then the other as the direction of the course was reversed. We found the difference was only two degrees[37].

Their experimental technique shows that they knew how to control for the effect of asymmetries in the air flow, that is, they could correct for this! Note how, with the apparatus mounted on the moving bicycle, their method would work best the lower the natural wind.

One thing they do not report is the changes they made in the position of the two test surfaces on the horizontal wheel relative to the source of the airstream: Comparing the two figures we see that in the proposed experiment, the airfoil is *upstream* of the axis of the wheel. In the report, it is mounted *downstream*. The first configuration would be unstable[38]. With the airfoil mounted downstream, the system is stable.

Why they do not explain the difference in the two figures is curious. Perhaps it reflects a characteristic of engineering thinking - namely to report only that which is the case now and true, not precedents that led you astray. Here in these historical sources we get a whiff of the engineer's discounting of history. Here too we see how, with remark made only of the proper function of a device, the possibility of instability remains an unknown to any new user. If the apparatus has the form shown in the second figure, stability is assured - "all other things being equal".

Another anomaly piqued their curiosity but one this time less easy to accommodate. They came to mistrust the constant coefficient attributed to Smeaton which appeared in the formula for the force on a flat plate. While their experiment showed what angle of attack was required to balance the plate, and hence gave the coefficient of lift as the ratio of the force on the airfoil to the force on the plate, if they took Smeaton's coefficient and calculated the *absolute* lift force on the plate, then, knowing the ratio from their experiments, calculated the lift force acting on the airfoil, they were surprised by how big the latter was. Though not part of the original experiment, this result led them dig deep where they had trusted before.

Thus provoked, they concocted some new experiments designed to compare the lift on a flat plate with the lift on a curved surface both inclined at the same angle to the air flow. Their words (including the figure) follow.

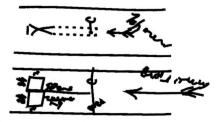

> In a square trough which served to keep the current straight, a wind vane mounted on an axis c was placed. The blades of the vane consisted of a plane 1" X 3.25" inclined to one side of the center and a curve 1"X3.25" inclined to the other side an equal amount. When exposed to the wind the vane took up a position to one side of the line of the wind direction thus showing that the curve required a less angle of incidence than the plane[39].

Again, the strategy is to explore what happens *in* the vicinity of an equilibrium state. If the force on the plane were the same as the (lift) force on the curve, the vane would *not* have moved aside in the air flow - the torque about the axis due to the lift acting on the plane would equal the torque about the axis due to the lift acting on the curved surface. If the device rotated, the surfaces moved to one side, the device would come to equilibrium when the torque due to the force on the plate balanced that due to the force on the curved surface.

A further manipulation of the apparatus, altering the angle of incident of one relative to the other until no angular deflection was observed, provided a basis for further comparison of their results with what others had published. "...I am now absolutely certain that Lilienthal's table is very seriously in error, but that the error is not so great as I had previously estimated..."[40]

Note again how the Wrights control for any asymmetry in the airflow ", errors which might otherwise result from variations in the force or direction of the wind at different places..." by running each test twice - the second time with the vanes flipped over so that the curved surface was positioned at the top of the trough and the plane at the bottom half. (The figure shows the original orientation). In the second instance, the vane would rotate opposite to that of the first instance so once again, averaging the two rotational deflections gave an unbiased measure of the advantage of the curved section over the plane section at the particular angle of incidence of the set up.

Their next variation on the theme is ambiguous in its representation. Clearly, they are controlling for asymmetry, but what
are they measuring?

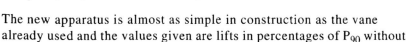

The arrow shows the direction of the wind. The dotted lines
show the orientation of the apparatus with the curved surface
oriented apparently at a negative angle of attack. (Lift, they
knew, was still possible in this case). The mirror image, solid
lines indicate that the apparatus was turned upside down in an
attempt to control for an air flow which was not truly in the
direction indicated by the arrow. Wilbur Wright explains further their purpose: (In this, P_{90} designates the force perpendicular to the plate, i.e., the "normal force").

> The new apparatus is almost as simple in construction as the vane
> already used and the values given are lifts in percentages of P_{90} without
> extended calculations. I think that with it a complete table from $0°$ to
> $30°$ can be made in thirty minutes, and that the results will be true
> within one percent. Errors due to variations in the position of the center
> of pressure are entirely eliminated. The same apparatus will also indi-
> cate the line to which the pressure is normal, so that the advantage of
> one surface over another in ratio of lift to drift can be obtained, and the
> truth of Lilienthal's tangential determined. We hope to have the appara-
> tus done within a week[41].

It appears that they are balancing the torque on a flat plate oriented normal or
perpendicular to the flow with the torque due to the lift of a curved surface. But
why the two arms extending out from the vertical axis of rotation? Extrapolating
back from their subsequent work, I conjecture a linkage meant to allow the curved
surface and the flat plate to move without rotation of either surface relative to the
wind. They had noted that "...an almost infinitesimal error is introduced by the fact
that the direction of the vane is inclined from one to two degrees to the direction of
the wind, first on one side and then on the other as the vane is turned over..."[42]
They reported how their new design enabled them to avoid the effect of shift in the
center of pressure - and, indeed, an analysis of their mechanism in accord with the
requirements of static equilibrium shows this to be the case: The angular displacement of the mechanism is independent of the location of the center of pressure.

But how did this device work? They claim that they will not only obtain the "...
lifts in percentages of P_{90} without extended calculations..." but will also be able to
deduce "...the line to which the pressure is normal..." An analysis, done in the
same spirit as the object-world analysis of the drift balance with which we started
this section, shows that measuring the angle of deflection gives the ratio of the Lift
force to the *sum* of the Drag force *and* the Normal force, the P_{90}, on the flat plate.
What the Wright's said they would obtain is the ratio of the Lift *alone* to the normal force. So something more must have been done.

I conjecture that they made two measurements, one with a plate sized to give a reasonable angular rotation of the device, a second with a smaller flat plate, say 50% the area of the first. This would give two states in which the Drag force and Lift force did not change but the value of P_{90}, the normal force, would. These two relationships could then be used to solve for the ratios of L/P_{90} and D/P_{90}.

This is a transitional apparatus: it is a design, yet to be realized, then re-shaped, re-thought, and re-used until the measurements they obtained satisfied their expectations.

A letter of October 16 shows the next edition; this time decipherable: It is very close to the one in which they had full trust.[43]

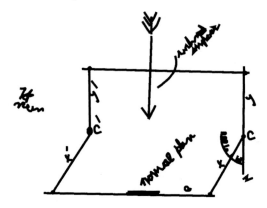

On the vertical axels cc' are fastened the horizontal arms x and x' which bear a crosspiece a on which a normal plane is mounted. The horizontal arms y and y' are equal in length to x and x' and are mounted on friction sleeves which fit the axles c & c'. They bear the crosspiece on which the surface to be tested is mounted. This method of mounting the surface preserves its exact angle of incidence regardless of the angular position of arms y and y', and also renders it indifferent where the center of pressure is located. In use, the surface is mounted on the cross arm at any desired angle and the wind turned on. The "lift" moves arms yy' to the right and the arms xx' which bear the normal plane to the left. The arms which have a friction mounting on the axels cc' are moved back to zero and readjusted till they remain there. The angle of the arms xx' is indicated by the stationary protractor. The sine of the angle zcs is the lift of the surface, for angle at which it is set, in percent of P_{90}.

A static analysis of the apparatus shows indeed that i) the sine of the angle indicated is equal to the ratio of the lift force to the normal force on the normal plane ii) the angle of incidence of the curved surface relative to the wind is independent of the orientation of the arms y and y' and iii) the result is independent of the location of the center of pressure. This last claim is counter-intuitive in that one might think that as the lift force vector moved out away from the axes of rotation, the torque would increase, throwing the system out of balance. But this is not the case.

Note too that, at this point they have done away with flipping the device over in order to account for changes in air speed and direction with location. Now they rely upon maintaining a unidirectional airflow by adopting an apparatus used by a Prof. Marey to (straighten) the flow and they enclosed all within a box with a glass cover - no longer needing to reach inside to draw on the bottom of the box, since the angular displacement would be reflected in the rotation of a pointer.

The final, archived representation of their instrument shows still further variations on the theme[44].

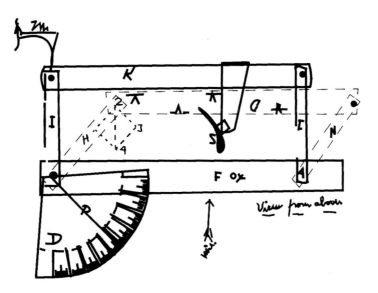

The two parallel links now are both located downstream of the airflow: This would diminish the effects of the curved surface - located upstream in the previous arrangement - upon the air flow impinging on and around the normal plane. The authors report in the text of the accompanying letter that they break up the normal plane into several flat surfaces of the same total area.

This last representation contains more information; it borders on what one might include in a patent application. It shows the apparatus in the before and after state - before adjusting the link carrying the curved surface to null out the effect of the drag force. It shows more details about the rest of the world; the scale, the reference, the supports. And it shows a diagram of the relative magnitude of the force component which acts on the normal plane and the lift force acting on the airfoil. (The dotted lines 2-3, 2-4 and 3-4).

Their drift measuring instrument is still another, and their final variation on this theme. But I stop my reconstruction here, not solely because of my felt need to move on but because the historical facts upon which any extended reconstruction would draw upon would have to include attention to the contribution of a contemporary, Spratt[45].

What knowledge is evident or constructed here? There is the production of trustworthy data connecting Lift and Drag to the angle of incidence of variously curved wing shapes. This information builds upon that originally produced by Lilienthal. It constitutes knowledge, but only when retrieved and used to provoke knowing anew (As in my reconstruction, in this immediate context, knowing about engineering knowledge).

Some scholars describe how, at first, the Wright Brothers complained sorely about Lilienthal's lift data, claiming it was seriously in error but then, once they had adopted a more appropriate value for "Smeaton's coefficient", discovered that Lilienthal's results were not too different from their own. This suggests to readers that if the Wrights had read Lilienthal correctly and had paid better attention to what others had used as Smeaton's number, they would have not had to go to all the trouble of conducting their own wind tunnel experiments. This would be a serious misreading of history, a mistake. Again it is important to distinguish between information and knowledge, or rather between information and knowing, both *knowing how* and *that*. By doing their own tests and working through the theory which governed the behavior of their instruments, in mind and in hand, they appropriated the relevant "information", made it their own and then some. Their production of this data provoked knowing about Lift, about Drag, about better airfoil designs, about the effective use of wind tunnels and the fundamental principles governing static equilibrium of rigid bodies.

All of this must be viewed within the context of the times. It is a mistake to take this data as we understand it today. Indeed, the way the Wright's themselves described it dis-allows so facile a reading. What is Lift? In what direction does it act? At one place they say it is perpendicular to the wind direction. In flight, this was not the same as the direction of the "natural wind". Their flight vehicles were glider's (and soaring birds) first, only powered airplanes second. There weren't any of the latter at the time. Yet, in the figure, taken from a letter to Chanute dated 5 January, 1902, Lift is shown perpendicular to the horizontal[46].

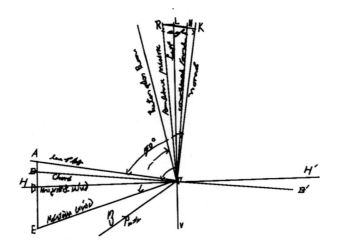

The Wright Brothers - and Lilienthal, Chanute, Langley and others - are not just doing experiments to measure the forces acting on variously shaped airfoils as a function of angle of incidence but are creating and constructing the language of flight - of lift, of drag, drift, chord, angle of attack - ab initio.

Throughout their efforts to build these instruments, we see them mix *knowing that* and *knowing how* - which suggests any attempt to disentangle Ryle's two forms, or to claim one is prior to another, would prove fruitless. The Wrights read texts - Lilienthal, Spratt, Chanute - and this provoked their critique. They read their machinery, responded to it as if they were in dialogue. They embedded their knowing (that and how) in their apparatus. This joining of craft know-how and scientific, abstract knowing together with this continual reconstruction of idea and artifact is characteristic of all thorough, modern, object-world development work. It is both challenge and reward in itself alone. We are not aware of the phenomenon because few engineers today document their work the way the Wright Brothers did, provoked by correspondence with a sympathetic and supportive flight enthusiast.

Before leaving this example, we critique the Nasa web page. Here the author's analysis appears rational and rigorous. If we accept his idealization of member D, shown in figure 4.4, his subsequent analysis does lead to the desired relationship between lift to drag ratio and the angle theta. The problem is that this idealization is incorrect.

We see this from my idealization of the top member: If, in figure 4.5, what I have called D_x and D_y are set equal to the drag and lift forces respectively as his enlargement of member D shows, then C_x and C_y must be zero for the requirement of force equilibrium of this top member to be satisfied. But then there is no way that for an arbitrary theta this member will be in an equilibrium state. For then we see that the force components Lift and Drag acting on the airfoil together with D_x and D_y, will, in general, conspire to produce a resultant torque on the top member; so moment equilibrium is not satisfied. Hence the equilibrium requirements are not satisfied. Our Nasa author's analysis is defective. Yet his result is correct! Only his initial representation, his isolation of the member D is in error.

In fact, this way of representing has a historical trace, starting with the Wright Brothers themselves. They show, in the figure included in the letter to Chanute dated Jan. 19, 1902, a schematic representation of forces very similar to NASA's picture and in the text of the letter explain:

> The lift on the surface S will swing the arms....only the lift exercises a twisting effect on the axles AA. The lift of the surface S is thus balanced against the normal pressure on the resistance surfaces RRRR.
> The dotted lines 2-3, 2-4, 3-4, show the resolution of forces. ...while 2-3 is the lift of the surface S as transmitted through K, I, B, A, H...[47]

Do we say, then, that the Wright Brother's analysis was in error? Or are we misreading their figure? After all, they don't present their figure as an isolated rigid body; the dotted lines are only claimed to show the relative magnitudes and directions of the lift force and the normal force on the plate. If their analysis was truly in error, if we want to claim that they didn't really have a full grasp of the concepts and principles of mechanics, then how did they obtain the (correct) result that the device would be insensitive to the location of the center of pressure? Here I run out of steam in my attempt, pace Collingwood, to replicate the Wright brothers knowing.

Reflections

While this last story is meant as history, like our recounting of bits of the work of our more respectable scientific persons, it too is a story of object-world work. And although the Wright's never read Navier's memoire - and Galileo never rode a bicycle as far as we know - the brothers are working within an object world which, at some fundamental level, is the same as the worlds of Galileo, Navier, and the authors of our modern engineering text; the world of requirements for static equilibrium; of force and torque, moment arms and resultants.[48]

These authors from these different historical contexts all speak the same language in this respect, but there are significant differences. The world of the Wrights appears as crude and awkward relative to our contemporary understanding of the application of the requirements for static equilibrium. Like primitive American folk art, the Wright's picture lacks depth. Still their vision is true to the needs of their task; after all, it worked as a basis for the design of their instruments. The world of Galileo has some of this same earthly flavor but there is a whole other purpose to Galileo's dialogue. It's not so much about how to build cantilever beams as it is about how natural philosophers might come to understand and explain, we would say scientifically, the workings of the most mundane phenomena. Navier's purpose too is theoretical as much as practical.

A comparison of the graphics of the three authors, and my elaborations (distortions) is useful: Galileo's beam as lever stays close to the physical object - by showing the supporting wall in such disarray he means perhaps to emphasize that we are not to see the wall as structure altogether. Only the letters intrude into our landscape. At the other extreme to Galileo is the figure from the today's engineering textbook. Here is total abstraction and generality; a specific instance of which could be taken as the same physical object as Galileo's cantilever but that is but scratching the surface. The Wright Brothers' sketches are intermediary: They suggest the physical apparatus but are just as concerned with the physical principles and concepts of force and moment, are just as much a display of these variables and their relationship, as are the texts of Navier and Galileo. Their sketches are essential to their work, serving both as a template for the construction of each device and to explain to others how their designs function.

Science provides a theoretical framework essential to object world work. But a distinction should be made between different kinds of scientific representations, and so different forms of information, which engineers draw upon in their work, in the writing of their textbooks, and in their designing. Cartwright makes the distinction between a fundamental mathematical theory meant to be universally applicable to the phenomena within a domain and phenomenological explanations which explain how cause and effect relate in particular phenomena[49]. Both are evident in this chapter. Navier's molecular force theory and Cauchy's continuum theory for the elastic behavior of a solid are instances of the former, of mathematical theories meant to be universally applicable. Galileo's analysis of the phenomenon of the fracture of a beam and the Wright brother's analyses of their instrument are instances of the latter. Contemporary textbooks in engineering mechanics of structures will include both forms of representation - presenting general mathematical theory and definition of concepts together with separate sections devoted to the explanation of the behavior of particular structural elements and systems - trusses, cables, beams and frames.

The effort made to bridge the general and the particular varies from one author to another, and often appears forced and inconsequential. But what is fundamental is not the integrity or completeness of the derivation of the particulars from the universal but that the students come to know the meaning of the conceptual entities and physical principles which enter into both forms of explanation. "To know the meaning" means to have some sense of the range of phenomena which might be so explained and how one constructs an explanation, develops ones own narrative and mathematical analyses when confronted with new phenomena, a new structural form, a new design. To know the meaning means to speak the language, to join the game, to know the rules and how and when they apply.

This orientation toward the pragmatic explains the openness of engineers to entertain and put to use different stories about, what some would claim, are closely related phenomena. Engineers are not, for the most part, so much interested in the development of unifying theories which might reconcile different ways of modeling one and the same structure in different contexts. There are those who are indeed interested in this; those who work at the development of general methods for the analysis of structures as in Finite Element Methods, but engineers in the main, out in the big world, make use of existing theory and methods in the explanation of how their alternative designs will behave in particular settings.

In this, they will generally make use of apparatus and prototypical hardware to verify the results of their analyses. The Wright Brothers development of instruments to measure the performance of airfoils of their design is an example. In working with prototypical hardware, the thinking through of how to make and remake it and the interpreting how it behaves is again done in terms of the same conceptual apparatus as is essential to reading the mathematical theory of elasticity or the engineering beam theory of Galileo. Prototypical hardware as well as general concepts, mathematical theory, equilibrium principles are all ingredients of the proper language of structural engineering.

Notes.

1. Duhem, P., *The Aim and Structure of Physical Theory*, Wiener, P., (trans) New York: Atheneum, 1962.

2. Duhem, thinks, according to Cartwright, that 'A physical theory is an abstract system whose aim is to summarize and to classify logically a group of experimental laws without aiming to explain these laws'. In engineering, their function is much the same, but now the group of laws are productive, prescriptive, structuring, not considered experimental.

3. Popper, K. R., *Objective Knowledge: An Evolutionary Approach*, Oxford, Clarendon Press, 1973, p. 60.

4. ibid, p. 73.

5. Couclelis, H., "Bridging Cognition and Knowledge", *Rethinking Knowledge: Reflections Across the Disciplines*, Goodman, R.F. & Fisher, W.R., (eds). Albany: SUNY Press, 1984.

6. Pirsig, *Zen and the Art of Motorcycle Maintenance*, New Your: William Morrow, 1984.

7. Wittgenstein, L., *Lectures on the Foundations of Mathematics: Cambridge 1939*, Diamond, C., (ed). Chicago: University of Chicago Press, 1976.

8. Gertz, C., *Thick Description: Toward an Interpretive Theory of Culture*, New York: Basic Books, 1973.

9. Searle, J., *Speech Acts: An Essay In The Philosophy Of Language*, Cambridge: Cambridge University Press, 1969.

10. Jammer, M. *Concepts of Force: A Study in the Foundations of Dynamics*, New York: Harper, 1962.

11. Theories change with aging. Quine sees the mix of theory and empirical "fact" as "...a man-made fabric which impinges on experience only along the edges..." Quine, W.O. "Two Dogmas of Empiricism", *From a Logical Point of View*, Cambridge MA, Harvard University Press, 1953. The core may remain sacred, but there can be much tinkering on the periphery which, in turn, is not without affect upon the interior.

12. Baird, D. "The Thing-y-ness of Things: Materiality and Spectrochemical Instrumentation, 1937-1955", *The Empirical Turn in the Philosophy of Technology*, Kroes, P., and Meijers, A., (eds), Elsevier, 2000.

13. This does not rule out the possibility that others, without the degree, can not think and know as the formally educated, as we shall see in what follows.

14 Timoshenko S., and Goodier, J.N., *Theory of Elasticity*, 2nd ed., New York: McGraw-Hill, 1951.

15 Todhunter J., and Pearson K. *A History of the Theory of Elasticity and of the Strength of Materials*, Cambridge: University Press, 1886.

16. Crandall, Dahl, & Lardner, *An Introduction to the Mechanics of Solids*, Second Edition, McGraw-Hill, 1978. p. 15.

17. I construct a reading of the text that goes beyond the intention of the authors in that my interest is broader. At the same time, I hope to construct a picture of the intention of the authors and of students' interpretations.

18. This is an example of a "control volume", a concept which Walter Vincenti sees as peculiar to engineering thinking.

19. In the latter half of the 18th century, various scholars postulated that the different properties of different materials derived from differences in the shape of their fundamental, constituent particles.

20. Note we are not talking here about the "actual" atomic structure of matter; that is another object world.

21 Truesdell, C., "Whence the Law of Moment of Momentum", *Essays in the History of Mechanics*, Springer-Verlag New York Inc., 1968.

22. There remains the other requirements, of compatibility of deformation and constitutive relations to be learned.

23. Navier, C.L.M.H., "Memoire sur les lois de l'equilibre et du mouvement des corps solides elastiques", *Mem. Acad. des Science*, 1824. Navier, under the influence of Lagrange, included no figure. This was not simply because the sparseness of the picture allows its dismissal and replacement by descriptive text, but because Lagrange's emphasis on mechanics as an analytical subject required no graphical representation to carry through a derivation or illustration of theory. That was his point.

24. Timoshenko and Goodier, op. cit. p 233. I present only one of three equations, the one representing force equilibrium in but one of three spatial dimensions, here indicated as "x". The other two equations have an identical form.

25. See Bucciarelli L.L., and Dworsky, N. *Sophie Germain: An essay in the history of the theory of elasticity*, Reidel, 1980.

26. The concept "force" has often disturbed the thinking of scientists: Besides its anthropomorphic leanings, so difficult to cast off, it has appeared to have mysterious qualities. These become less bothersome, less prickly, the more one invests in mathematical representation and less in physical meaning, dropping questions prompted by common notions of push and pull. Heinrich Hertz would have liked to have eliminated the concept altogether. Hertz, H., "Introduction", *The Principles of Mechanics*, New York, Dover Publications, 1956 (1895).

27. Again, this is a corruption of the text; Galileo writes no equations in this form. Galileo Galilei, *A Dialogue Concerning Two New Sciences*, Crew, H., and de Salvio, A. (trans) New York: Macmillan, 1914.

28. Photograph courtesy of the Archives of the Smithsonian Institution, National Air and Space Museum. Negative number 2002-16631.

29. http://quest.arc.nasa.gov/aero/wright/team/fjournals/hange/balance.html

30. Ryle, G., "Knowing how and knowing that", *Proceedings of the Aristotelian Society*, vol. XLVI, 1948.

31. It's tempting to label the second context, the historical context, a *context of discovery* where *knowing how* dominates. The first, a *context of verification* where *knowing that* is put to the test.

32. Agassi, J. *Towards an Historiography of Science.* s'Gravenhage: Mouton & Co. 1963, p. 74.

33. Collingwood, R.G., *The Idea of History.* Oxford University Press, 1946, p.241.

34. *The Papers of Wilbur and Orville Wright*, MacFarland, M.W. (ed) McGraw-Hill, 1953.

35. ibid., p.120.

36. This point seems to have been missed by most historians. For a full account of the Wright Brother's contribution to the development of the science of aerodynamics see Anderson, J.D., Jr. *A History of Aerodynamics and Its Impact on Flying Machines* Cambridge University Press, 1997. Anderson is one of the few who get this right.

37. *The Papers...*, op. cit., p. 125.

38. Consider the first sketch of their apparatus: Say the torques did balance, i.e., the lift on the curved surface was equal to that required for equilibrium. If a disturbance set the wheel to move clockwise, the angle of attack would increase, the lift would increase and the torques would no longer balance, the wheel would rotate clockwise still further, further aggravating the imbalance. If a disturbance sets the wheel going the other way, the lift would decrease, the wheel would rotate counter-clockwise still further, so again the setup is unstable. (This assumes the relative change in the magnitude of the force on the flat plat is less than that on the airfoil for the same angular deviation from the equilibrium orientation)

39. *The Papers...*, op. cit., p. 125.

40. ibid., p. 127. They came to realize that the value for Smeaton's coefficient they had adopted was "incorrect", not the best available according to most contemporaries.

41. ibid., p. 127.

42. ibid., p. 126.

43. ibid., p. 134.

44. ibid., p. 206.

45. For clues to the role of Spratt in the development of the wind tunnel balances see *Miracle at Kitty Hawk, The Letters of Wilbur and Orville Wright*, Kelly, F.C., (ed) New York: Farrar, Straus and Young as well as MacFarland, op. cit.

46. *The Papers...*, op. cit., p. 199.

47. ibid., p. 204.

48. Some claim the Wright brothers were endowed with the capacity to do "visual thinking". The attribution of special mental capacities strikes me as unnecessary. Claims like this tend to bring any analysis to an end, *tout court*.

49. Cartwright, N., *How the Laws of Physics Lie*, Clarendon Press, Oxford, 1983.

5

Learning Engineering

The past decade has seen a surge of significant activity aimed at the renovation of engineering education. In this chapter I review some of the changes proposed, attempted and implemented, focusing on developments in the United States. I intend to analyze these developments in broad terms, testing to see if a critical, philosophical perspective can help us better understand both the fundamental notions motivating and framing change and the likelihood that the proposals for renovation might work – and if not, what new ways of thinking about "the problem" might do better. I am particularly interested in how certain deficiencies and lacunae in the way engineers see the educational process stand in the way of needed reform. First an overview, a bird's eye-view of the traditional system.

Engineering education in the US is the business of universities in the main. The major state universities have colleges or schools of engineering. Other private institutions also have strong engineering programs - MIT, Cal Tech, Carnegie Mellon, Cornell, Rose-Hulman, Harvey-Mudd, Lehigh, Dartmouth. Not all institutions offering engineering programs carry on in the same way or are of the same scale; for example, not all qualify as research universities, though most aspire to be recognized as such.

Undergraduate degree programs across this range of institutions have certain common characteristics. At MIT, as at most universities and schools of engineering, the curriculum requires a thorough grounding in the basic sciences and mathematics during the first year or two, several courses in the engineering sciences relevant to the student's chosen discipline, one or several laboratory experiences, more specialized advanced courses in a subdiscipline, then a capstone design course, thesis, or some type of project work in the fourth and final year. The students must also spend a significant portion of their time in the Humanities, Arts and Social Sciences. Some programs deviate in significant ways from the norm, but in the main, from a bird's eye-view, this is what one sees.

Students are mostly male; but recent government funded programs and university leadership have had some success in attracting more women and minorities into the engineering profession. Our youth enter university at the age of 17 or 18, and are expected to complete a degree program within four or five years. There are part-time students, mature students, but the norm is as I have reported it.

One often neglected student characteristic in planning new curricula is what students intend to do upon leaving the academy. Faculty generally assume that their graduates will pursue an engineering career, broadly conceived to include management and postgraduate study in engineering as well as work within an engineering firm. While the university may have orientation programs and offer advice to those who seek to do otherwise, e.g., to go on to medical school, to law school, or even to prepare for secondary school teaching, the core curricula are defined by departments, of Mechanical Engineering, Civil Engineering, Chemical Engineering, Electrical Engineering...etc. and are set by faculty in accord with the expectation that the student will become one of them.

Faculty are predominately male; diversification here has proven more difficult, despite the inducements offered through government programs and the implementation of new policies of enlightened university administrators. To be hired onto the faculty requires, in all but the most exceptional instances, a Ph.D. degree, evidence of research potential and an interest in teaching. Traditionally (the past 50 years) hiring, like curriculum planning, has been the business of departments with slots distributed over the different divisions within the department. This may be changing with changes in the way funding of research is managed and controlled. Central administration is gaining more control as the federal government increases funding of research centers which focus on "hot" topics deemed important to the security and welfare of the nation. These research questions are generally interdisciplinary in nature and so require the integration of the efforts of faculty from distinctly different disciplines and departments. As provosts and presidents have worked to meet this demand, they have in the process gained more of a say about the appointment of faculty.[1]

The best and most of the faculty openings are "tenure track" positions. A freshly minted Ph.D. fortunate enough to obtain such a position will immediately become an Assistant Professor. Generally he or she has seven or eight years to prove worthy of tenure - which still means job security for life. Research production remains the most important criterion in this respect. Promotion to Associate Professor follows, although some departments will allow promotion to this rank before a tenure decision need be made. Tenured, Associate Professors generally can look forward to becoming Professors as long as they continue to produce quality research, serve the profession and the institution in expected ways, and teach when called upon.

The specific subject matter of research will depend upon the focus of the faculty member's department, division, or research center as well as upon the individual's field of expertise. But even within a specific domain, e.g., engineering mechanics, research takes different forms. Some individuals will appear to behave as the most theoretical of scientists, seeking a better understanding of particular phenomenon relevant to their domain; e.g., those that develop mathematical/theoretical explanations of crack propagation in structural materials. Others will do laboratory experiments to verify new theory so constructed. Laboratory testing of a phenomenological conjecture - e.g., to test the efficacy of surface reinforcement of concrete beams - is another respected type of research.

Others will rely heavily upon mathematics but their interest will be in the development of efficient methods for the evaluation of the consequences of theory when applied to particular problems and in engineering design. The development of finite element methods for the calculation of the stresses and deflections of particular structures engaged an international research community of faculty over the past half century.

Still others do research on "systems" rather than particular physical phenomenon. Faculty in mechanical engineering do research on manufacturing systems, in aerospace engineering, on air traffic control systems, in chemical engineering, on processing systems; in civil engineering on construction systems, water resources, and environmental systems. Research on methods for multi-objective optimization of systems would fit in this category.

Finally there are those who must contend with social and political features of systems in more explicit ways, e.g, in research programs on technology and policy. Interdisciplinary research projects, in bioengineering, have a still different, mixed form.

The ideology of engineering research is the ideology of the science laboratory. This is not to claim that the stuff, the content engineers confront in "the lab" is very much like the content of the scientists' explorations. Indeed, there may not be any stuff at all, or what is seen and manipulated has but a virtual presence. Rather, I mean the nature of the intellectual enterprise is the same: Boundaries around the research task or program are drawn close in, as in physics, chemistry, mathematics. The "all other things "of the *ceteris paribus* clause make up a very large set. A piece of the world is defined, reduced down out of the big world, so that questions relevant to the field can be well posed and quantitative results obtained in response. The world of the engineering research task is an object world *par excellence*, even more narrow and limited than the corresponding disciplinary matrix essential to the work of the engineer out in the world, participating in the design and development of a new product.

I speak of engineering research though my topic is engineering undergraduate education because I want to explore the fundamental beliefs and values teachers of engineering bring into the classroom and how these influence educational reform. The way we see things as teachers - the world, our students, our course content, concepts, principles and worthy methods - depends upon the way we see and carry out our research. Our ideas about what constitutes an important and well posed research question, what ways of thinking are rational, what methods are appropriately applied, and what will constitute a significant result – all of this guides and constrains our educational efforts, sets our objectives, fixes course content, defines our relationship with students and how we measure their progress. My claim is that the scientific and instrumental essence and ethos of research fixes and limits the way we see and attempt to reconstruct undergraduate engineering education.

Research ideals and norms undergird the following propositions which, I claimed at one time, set the framework for engineering faculty thinking about engineering education[2]:

- Mathematics and science are primary. Instrumental methods, and the reductionist perspective at their core, are essential to professional practice.

- The technical problems engineers confront have technical solutions; unique solutions found using the instrumental methods learned in school.

- Knowledge is like a material substance. It can be segregated into independent blocks. Curriculum design is a matter of choosing the right blocks and arranging them in the right sequence.

- Engineering faculty are the authoritarian sources and dispensers of all knowledge; we transfer the blocks to our note-taking, receptive students.

- Our graduates will enter engineering practice; there they will be rewarded for their effectiveness and creativity in applying their blocks of knowledge as individuals, working alone.

- What we teach is value free, context independent, and universally applicable.

Over the past few decades, changes have been made in engineering programs which indicate that these propositions are no longer taken as definitive. Change has been motivated in part by evolving images of what faculty see their students doing when they enter the profession.

When I put forward the postulates above, I wrote also of two images of the practicing engineer – one traditional, the other more modern. I did this because I wanted to include in my analysis of curriculum reform the "needs of industry" as well as research ideals and norms because faculty do speak of the former as well as the latter when discussing curriculum reform. These two representations are to be taken as ideal types, as strawmen if you like, or stereotypes, of our typical graduate at work; they are two different visions we, as engineering faculty, carry around with us and reference at curriculum discussions, committee meetings and the like. I labeled the traditional type the ideal of the 50's. Another is less clear but forming; I labeled it the ideal type of the 90's.

The 50's type is well prepared in the sciences[3]. He applies this knowledge to complex, high-tech problems of the kind encountered in the engineering of complex military-industrial and aerospace systems. He does this as a staff employee within a large, well equipped, well organized and authoritarian organization; he toils within a bureau defined by its technical focus and expertise.

His company allocates a significant portion of its budget to research; a strong research base is essential to its development of sophisticated "closed systems" - products and systems designed for the benefit of all mankind, many for the security of the nation. After a few years our 50's type may return to the university to enhance his research skills perhaps to pursue a Ph.D. then either re-enter the industrial or government laboratory or join the faculty of some ever-expanding engineering school. While those who choose to remain in industry will eventually take on managerial responsibilities – as much as they disavow interest in this aspect of the business – there remains the possibility of dual career ladders to ensure that technical expertise receives its just rewards.

The ideal type of the 90's is of a different sort. She is also well prepared in the fundamentals, but now of a more diverse set of disciplines. Prepared too to work in teams; able to articulate, communicate, and defend a proposal. The 90's type is open to negotiation, knows how to cope with uncertainty. She finds her work varied and multifaceted - negotiating a product specification with marketing on Monday, the next day finding time to take some product reliability data down in the lab. She seems never to have enough time to meet the demands of the three projects which require her talents. Her firm, the firm of the 90's type, is of smaller scale - profit driven, thriving, or trying to survive, apart from large government contracts. Its products are information technology-intensive yet designed for the general public. Product design and development is now a favored activity; entrepreneurial activity likely.[4] When the 90's type returns from this hectic world to the university, she is as likely to take up a program in business, or in technology and policy, as much as in engineering. Law school is a possibility as well.

If we see our graduates as 50's-type, then there remains no dissonance between the aims of research and the aims of education. If, on the other hand, we picture them as 90's-type, then we have a disjunction between the aims and ideology of research and of education. My claim is that we are moving from the old to the new, for there is strong evidence that faculty and schools and colleges of engineering are energetically developing new ways of thinking about what we want our students to understand and be able to do - if our program of study is to be effective and meet the needs of the times – as well as developing new content and means for its teaching. I turn to review some of this activity.

Reform

One need only turn to the journals of engineering education first to note their increasing number, then the range and tenor of the articles, to see the growth in interest in the reform of undergraduate engineering education over the past decade on the part of faculty. The National Science Foundation has funded programs aimed at reform at levels hitherto unseen; and ABET, the agency responsible for setting accreditation criteria, has recently revised its recommendations for undergraduate engineering degree program content and implemented a whole new system for evaluation of the same.

Professor Richard Felder et. al. have summarized the Accreditation Board for Engineering and Technology's new requirements:

> ...we must strengthen our coverage of fundamentals; teach more about "real-world" engineering design and operations, including quality management; cover more material in frontier areas of engineering; offer more and better instruction in both oral and written communication skills and teamwork skills; provide training in critical and creative thinking skills and problem-solving methods; produce graduates who are conversant with engineering ethics and the connections between technology and society; and reduce the number of hours in the engineering curriculum so that the average student can complete it in four years.

The authors go on to note that

> ...even if nothing new is added to the existing curriculum, confining it
> to four years will be almost impossible unless more efficient and effec-
> tive ways to cover the material can be found[5].

This daunting challenge has not been shrugged off. Faculty and departments,
Deans and whole institutions have invested heavily in reform. Surveying the different school and university programs, we see a variety of proposed and actual renovation but, at the same time, only a handful of primary themes.

One common recommendation is to shift from passive to active learning. Felder
et. al. again:

> In the traditional approach to higher education, the professor dispenses
> wisdom in the classroom and the students passively absorb it. Research
> indicates that this mode of instruction can be effective for presenting
> large bodies of factual information that can be memorized and recalled
> in the short term. If the objective is to facilitate long-term retention of
> information, however, or to help the students develop or improve their
> problem-solving or thinking skills or to stimulate their interest in a
> subject and motivate them to take a deeper approach to studying it,
> instruction that involves students actively has consistently been found
> more effective than straight lecturing...

> ...the challenge is to involve most or all of the students in productive
> activities without sacrificing important course content or losing control
> of the class[6].

As an example of the shift from passive to active learning, some engineering
programs have taken the "introduction to engineering" course taught early on in
the student's career, dropped the lecture format, and actively engaged students in
design projects over the semester. While requiring no sacrifice of important course
content, this innovation has none the less met resistance as it calls into question
the traditional notion that students can't design until they have studied and mas-
tered the content of the different disciplines - the really important fundamentals -
which must be called upon in the creative synthesis of new products and systems.
In Mechanical Engineering, for example, the claim is made that one must study
structural mechanics, fluid dynamics, controls and thermodynamics before one can
do serious design work, of say an automobile. This is why the study of design traditionally has been left to the senior year, to a "capstone" design course. But now
many schools and colleges of engineering require first year students to do design.
These are not "drawing" courses but meant to teach some of the engineering fundamentals within the context of a design task - e.g., design a playground ride, a
windmill, rope bridge, and the like. The notion that science and theory go first is
being subtly challenged. The new idea is that one can go directly to more phenomenological understandings and do real engineering work in any field.

Other reform goes under the banner of "hands-on". Here there is explicit recognition that science learning alone does not suffice to survive, none the less excel,

in engineering practice. There is a "back to the basics" air about the argument for the re-introduction of laboratory work, machining and hardware manipulation skills, design-and-build projects, prototyping and testing – but it's a different kind of basics than Newton's laws or the mean-value theorem. Product dissection, where the students take apart a mechanical contrivance and attempt to figure out why it has the properties it has, is one way of engaging students in the discovery of engineering principles and appreciation of technique and craft.

The emphasis on active learning is also what justifies proposals for 'problem based learning' and the integration of open-ended, design type exercises throughout the curriculum, not just in the first or the final year. Some faculty, including your author, have promoted the introduction of such open-ended exercise into all courses - engineering science subjects, laboratory courses as well as design courses – at all levels. "Open ended" means that there is more than one answer to a problem, more that one method to apply in getting an answer in many cases, and that the student must participate in formulating the problem in the first place. The aim is not to teach design per se but rather to redirect learning toward certain features of engineering professional thought ordinarily discounted, if not simply ignored, in the traditional curriculum and which are essential to critical, reflective practice.

Probably the least successful attempts at reform have been those which have attempted to capitalize on the ever-increasing power and sophistication of computer and information processing technique but have failed to take full advantage of the technology. If one tries to incorporate the new with the old without changing the old - that is, without considering changing the course content and what we think is fundamental that the student learn - the results are bound to fall short of expectations. If our assignments remain limited to well-posed problems with unique solutions, then the use of a computer is hardly justified.

The failure of the technology to live up to its promise may also, paradoxically, be blamed on the innovation itself, i.e., on the rate at which the technology is improving each year. Faculty generally do not change the resources they use in a course over a four or five year period. The syllabus, a favored textbook, a library of problems and exercises, are relatively stable resources. In the same period of time, changes in IT capabilities have been, and look to continue to be, dramatic. Given this, faculty are not likely to invest time and energy, none the less funds, in adopting and adapting computer technique and technology if it will have to be redone again in the next year or two. Still, each year sees further incorporation of the technology in the lives of students and faculty, e.g., in outside-the-classroom, ad-hoc use as well as in applications in conjunction with substantial renovations in course content and purpose, e.g., the infusion of open-ended exercises throughout the curriculum.

The above observations applies to traditional courses. There is one additional very important way in which IT *is* changing the landscape: The technology itself has become a thematic line in curriculum reform. That is, computer and information technology is seen to so dominate the development of new products and systems – whether pharmaceuticals and their processing, air-transportation communication and control systems, toys for children, toys for adults, even "smart" civil engineering structures which respond to the environment or call home when something goes awry – that major shifts are in the making within engineering departments to engage the challenges of designing and building these systems. We see the introduction of new courses and even "tracks" within the departmental major focused on information systems. Within these new programs there is a double opportunity for change. The fundamentals and what's to be studied are in construction and this, together with the newness of the technology itself, presents a relatively unconstrained field for trying out new ways of teaching and for fostering new ways of learning. Time will tell how this challenge is met.

The essential ingredient in all of the more successful attempts at reform is a move toward active learning and a refocusing on the student experience. We see a shift from the didactics of single answer problems, which traditionally has been the steady diet of most engineering courses and "canned" laboratory exercises to open-ended tasks. Knowledge becomes, not a packaged commodity, but an event.

Critique

While the reform of undergraduate engineering education is serious and better funded than perhaps at any time in its history, there are certain features of the effort which call for critique. Yes, we faculty recognize the virtues of actively engaging the student in the classroom; no longer limit our deliberations to questions about what topics and methods to include in our syllabus; now experiment with innovations in the classroom that appear to shift the focus onto the student and their experiences in learning; and even take assessment – of our innovations, of our teaching, as well as of our students – more seriously. Still, our discourse is constrained. When we look at the challenge of reform, we see it through the tinted glasses of instrumental, scientific analysis. This, as I hope to demonstrate, as already evident in Felder's laying out of the challenge of reform, seriously limits possibilities for renewal.

Take, for example, the proposal Felder advances for how we should set instructional objectives and test to see if our students measure up:

> The behavior specified in an instructional objective must be directly observable by the instructor and should be as specific and unambiguous as possible. For this reason, verbs like *know, learn, understand,* and *appreciate* are unacceptable.[7]

What students should be able to do is: *define, calculate, estimate, outline, list, identify, explain, predict, model, derive, compare and contrast, design, create, select, optimize* - which presumably are "directly observable". While he allows that bringing our students to know, learn, understand, and appreciate "... are critically important goals,... they are not directly observable". What's implied is that testing whether students can define, calculate...etc.will provide an indirect measure of whether or not one is meeting these other, what many faculty would call softer[8], educational objectives.

Here now we focus on the student, but it is a student seen as a behaviorist's object; a creature fed with instructional materials and which, upon prompting with the right sort of query, outputs a response which we directly observe and measure. In the light of this recommendation, the meaning of the call to shift our focus from *teaching* to *student learning* becomes ambiguous and problematic.

This same instrumental perspective is reflected in the metaphors we use when explaining what we are about. I have already described the "knowledge as stuff" metaphor with its implication that more is better. It is there again in Felder's evaluation of the challenge set by ABET when he notes the impossibility of fitting everything, old and new, into a curriculum of but four years duration. It is reflected too in the way faculty propose to include new content – e.g., ethics, communications skills – by adding entirely new courses.

The closely allied metaphor of "mind as computer" is evident in the behaviorist's imagery of Felder. More explicitly he writes:

> Our goal in teaching is to get information and skills encoded in our students' long-term memories.[9]

Another common metaphor used in talk about course requirements and degree programs is what I call the "production metaphor": Here faculty picture their students as "products for industry". Their vision is of a mass production process; their job to take hold of students when they enter the department, then process these crude, raw materials and shape and polish them as they move along, lock step, through the curriculum. In the end, if not rejected, they expect their students will be fully functional and fit to meet "the needs of industry".

This same metaphor is evident in the proposals made by faculty who advocate curriculum reform based upon "just in time" learning. Usually this method is invoked in the context of project based, case study or design task learning. The idea is that we expose and teach the student only those materials and at a level which is appropriate to addressing the immediate exercise. Faculty bring the knowledge components and resources into play at the time when they are needed, delivering what is required to solve the immediate problem and no more.

To its credit, the method emphasizes the importance of synthesis relative to analysis and, as such, can indeed promote active learning. All of this is well and good - in Felder's terms, efficient and effective. The problem is that this way of speaking and thinking about the students has consequences that are not so healthy. For example: Objective testing, under the production metaphor, quite naturally leads to measuring a student's achievement by their location in a bell-shaped frequency distribution. Faculty talk about student quality and worry about whether their educational standards are too loose or too tight. Some departments, in accord with the dictates of statistical quality control, may even require that a certain percentage of students, e.g., those lying at the left hand tail of the distribution, should fail. This is not to say that representation of students' learning in this aggregate way is not useful and informative. But does it inform us about student achievement or is it a test of our effectiveness as teacher? Perhaps we should inspect the machinery of production.

All of this imagery reflects an instrumental view of the student as object, as artifact, as computer, as product, as bucket, as container to be filled or charged, loaded with the stuff of knowledge, and implanted with appropriate skills. It is a way of seeing consistent with the kinds of representations engineers construct as researchers in their modeling of systems with users as ergonomic appendages or as aggregate consumers with certain limited preferences. If we insist on keeping our tinted glasses on, it is very difficult to even conceive of how it might be different. If we don't take them off, it will be impossible indeed to meet the challenge ABET has set. There just isn't enough time or space in the four years to "cover" all that needs to be covered.

An alternate way of seeing

Israel Sheffler, in a collection of essays in the philosophy of education, in a chapter on teaching basic mathematical skills, sees this world differently. He claims that it is not sufficient to take the concept of knowing as the acquisition of the distinctive principles and methods of a subject as the sole basis for setting instructional objectives and curriculum requirements:

> The aims of education must encompass also the formation of habits of judgment and the development of character, the elevation of standards, the facilitation of understanding, the development of taste and discrimination, the stimulation of curiosity and wonder, the fostering of style and a sense of beauty, the growth of a thirst for new ideas and visions of the yet unknown.[10]

> ...successful performance in mathematics rests not only on general skills but also on general attitudes and traits such as perseverance, self-confidence, willingness to try out a hunch, appreciation for exactness, and still others.[11]

And in another chapter in which he analyzes and exposes the shortcomings of the computer-metaphor in speaking of mind, he lays out, contra Felder, why the "...concept of information is far from capable of adequately expressing our educational aims."

> Even the capacity for *intelligent use* of information will not suffice to express our educational aims relative to problem-solving...

> Problem-solving, further, needs not just the recognition and retention of facts but the recognition and retention of difficulties, incongruities, and anomalies. It does not simply affirm truths but entertains suppositions, rejects the accepted, conceives the possible, elaborates the doubtful or false, questions the familiar, guesses at the imaginable, improvises the unheard-of. An intelligence capable only of storing and applying truths would be profoundly incapacitated for the solving of problems.[12]

Here now is an alternative perspective. It is so different that we find impossible to see how it might have relevance to engineering education. Rubbing our eyes, we explore and try to envision how it might prove enlightening.

We don't have to buy the whole package. We can leave the development of character and the elevation of standards to the side for the moment. On the other hand, this talk about curiosity and wonder, new ideas, the yet unknown, and unheard-of sounds much like the kind of talk we would engage if our topic were how to foster creativity and innovation. And we have already reported on how practicing engineers spend a good bit of their time and energy in striving to explain why their productions malfunction. Problem solving in this mode indeed requires the identification of difficulties, incongruities and anomalies, the entertaining of suppositions, the conception of the possible, etc. More problematic, perhaps, is the idea that we elaborate the false or even reject the accepted.

But let us accept Scheffler's broad sweep of the ingredients as pertinent to the education of engineers; let us acknowledge that we seek to do more than convey to our students information about concepts and principles, laws and properties, and algorithmic methods for solving problems. Indeed, we may already claim that we do attend to these, at least implicitly in our lectures and exchanges with the student. How might we proceed?

The first step is to explicitly recognize that process is as important as product; that means are as important as ends; that the ways the exercise is set, the ways the students muddle through, are as important to address as are the answers they submit. We broaden the scope of "fundamentals" to include the recognition of difficulties, incongruities, and anomalies; the entertainment of off-the-wall (perhaps) suppositons; allow questioning the familiar, encourage improvisation et al.

The next step is to imagine how we might actually do this, that is, construct topics and exercises which would serve as the medium for teaching/learning the fundamentals, now encompassing more than those at the focal point of the traditional content of our course. In an engineering mechanics course, the fundamental concepts and principles which describe and explain the behavior of structures and machines remain very much at center stage; Force and displacement, equilibrium and continuity, friction, elasticity, plasticity, momentum and energy, statics and dynamics remains the proper language to learn. The aim is not to displace the traditional but rather to enrich the context of their learning.

I consider three approaches; the first makes use of history; the second relies upon student misconceptions; the third, open-ended, design type exercises of the sort already described.

Using History.

History, the history of science and technique, is full of incongruities, suppositions, rejection of the accepted, guesses at the imaginable and improvisation. The serious study of history can be revealing and a contribution to knowing, even when our predecessor was in error. Take, for example, Galileo's faulty analysis of the beam as a lever. In our commentary we described how, though his result was in error, his method was thoroughly modern. Indeed, that's what makes Galileo a pivotal intellect in the rebirth of science, a giant upon whose shoulders we stand. And it was as much a technical event as it was scientific.

He shows us about the relation of abstract model to concrete artifact; he relates cause to effect in an analytical narrative; he explains the purpose of his analysis, how it can be used to predict when a cantilever beam will fail; he reveals the power of abstract representation, of underlying form, in describing how a whole class of structures will behave; he attends to scaling up and scaling down in that his analysis applies to the bones of giants as well as needles.

That he was in error can prompt other questions relevant to our new range of fundamentals. We can "correct" his vision; we can draw an abstract lever, like the one I overlaid on his figure, only this time remove the fulcrum and make it a true isolated body to which we apply Newton's laws. We can then engage our students in conjecturing and with suppositions of a different sort: i.e., why did Galileo go wrong? Why didn't he recognize that equilibrium of forces was not satisfied? This line of questioning can serve as a powerful vehicle for bringing students to speak the language of "free-body diagram", of the fundamental importance of a proper isolation in the application of Newton's Laws. At the same time it can be a confidence builder - to recognize that Galileo had the same difficulties as they do in constructing an idealization given the "real" beam.

This last remark points to still another line of questioning: What does it mean to say he was in error? Indeed, how justified are we in projecting back onto the first few decades of the seventeenth century, our ways of seeing, of speaking, of representing, of analysis? Here now is a whole other kettle of fish. At this point, if we find ourselves so engaged and the student expresses interest in how historians meet this challenge, it is perhaps best to recommend he enroll in a course in the history of science. But note that the task of diagnosing error is akin in many respects to the task we face in explaining the failure of modern technology - a task that requires a social consensus - in the case of Galileo, of historians of science.

Navier's modeling of an elastic solid as a collection of molecules which interact, exerting sensible forces at insensible distances, can also be usefully treated in a mechanics course. Here the question of the relation of the abstract to the concrete looms large. This is as much a philosophical question as it is historical. But the history makes the question real: The world of inter particulate forces is a possible world which was indeed a world inhabited by mechanicians of the nineteenth century. We can there engage in an exploration of the ontology of models: How "real" are they? Is the claim that the solid behaves "as if" it were made of molecules so acting but not that this is a true picture of the material at some atomic level? At another level, the question is pragmatic and relevant to the challenges engineers face in today's world: How crude, how sophisticated a model do I need? In what ways might my representation be deficient?

Navier's model is a more natural representation of a solid, elastic body than that which speaks of stress and strain at any point in a continuum, as Cauchy soon enough did. That is, it follows naturally from Newtonian ideas about the behavior of isolated particles subject to external forces. Our textbook authors followed this tack in deriving the equilibrium requirements for a rigid body. Considering Navier's way of seeing before Cauchy's construction then provides a more gentle sloping path up to the concepts of stress, strain et al. (Do students repeat the mistakes of the past?)

The Wright brothers, in their testing of different airfoil shapes is another historical resource: The development of the balance could serve as an excellent example of the power of abstraction in the analysis of the static equilibrium of rigid bodies as well as an introduction to the concepts of lift and drag. Students would be challenged to construct the appropriate idealizations of all the members, introducing a complete and consistent set of internal force components, then deduce a set of relationships which ensure that the requirements for static equilibrium are met for each isolated member. They could critique the Wright's method of analysis as reported in their letters to Chanute, conjecturing the missing steps, imagining what resources, both intellectual and material, they had to draw upon. They might be challenged to actually replicate the apparatus and the tests the Wrights themselves carried through.

There is much here that faculty would agree is valuable for our students to learn, to experience, using history. But there are also problems with this approach, some of which we have already alluded to. For instance, to deal with history in anyway other than superficially requires reaching beyond the myths and digging into source materials first hand. While there are some secondary sources that will serve in most domains of the sciences relevant to engineering, these generally are not written for a student. That the history of science is not always what it has been made out to be is evident in perusing the *Structures of Scientific Revolutions*: Kuhn's famous treatise can be read as a strident critique of the disregard and distortion of history on the part of teachers of science - at all levels. One of the norms of the profession appears to be to disregard the past, to continually repaint the picture of the achievements of the past in accord with how we see the world today. Another well known historian of science has gone so far to suggest that engaging physics majors in serious study of the history of their chosen field might be dysfunctional.[13] To assure that *using history* will not become *abusing history* will require faculty preparation of a different kind.

A more subtle danger: In any honest enquiry into the work of a Galileo, a Navier, the Wrights or even a Laplace or a Newton we might wonder if these persons were deficient in some way; was their mental machinery defective? Or does the dissonance between the old and the new have more to do with our inability to see that what we claim as facts, true concepts and principles, may be historically contingent and contextually dependent? This is a whole other Pandora's box actively engaging scholars in the study of science, technology and society. If history is to be used, we must attend to this dimension; otherwise we shortchange our students in shielding them from the importance of social context in the development of science and technology.

When to cut off the discussion, how much time to spend on a particular historical episode, is another question that has to be faced. For that matter, where in the syllabus will we find the time to do all of this? And remember Felder's concern: to engage in these kinds of explorations is liable to end in loss of control. Who knows where the probing of the historical sources might lead? What if the student really develops an interest in the challenge of doing history, or won't drop the discussion until he has figured out how the Wrights actually thought through their design of the balance?

We must resist these thoughts; we must leave our tinted glasses on the table. We must free ourselves from the image of a course as containing quantities of knowledge which we must deliver to our students. Focus on "attitude" and "perspective" rather than what content to cover. And as for loss of control, control over what? Is it over the amount of material we want to cover? Or is it more the fear that the student will ask a question to which one does not have a ready answer? Or that another student might chime in with a better one than the one you hastily constructed? If the latter, this itself can be an effective educational strategy, i.e., to allow students to join in the development of ideas in a more free exchange than is possible within the one-way dialogue of the traditional classroom - a more honest Socratic technique.

Using students' misconceptions.

In using history, we study ideas in a world of the past, of past institutions and infrastructure, of peoples and apparatus long since gone, the evidence for which survives only in texts and an occasional artifact. There we find incongruities, anomalies, concepts we might never imagine, often error and even the still unheard of. But we don't have to reach back to the past to encounter these phenomena in rational scientific thought; we need only listen carefully to our students.

The notion here is that we can bring those new to our field to learn from reflection upon their erroneous ways of thinking about the world. These ways of thinking are often labeled misconceptions; they are to be treated, not as something to be washed away or repressed without notice but as stories, like histories, to be analyzed in hopes that we can decipher and discover why anyone should hold to falsity. If successful in this, we can rectify our student's ways.

I want to speak more about these supposed cognitive malfunctions for I find in this topic, as in history, a way to provoke new thinking of what it means to know, to understand, to learn in engineering as well as in science - a way to again address our renewed fundamentals for problem solving. I use as a vehicle for this exploration an 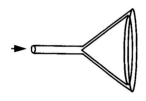 exercise found in a physics textbook and conjecture how a student might approach, reason, imagine and finally be freed of his misconception.

> Explain why you can't remove the filter paper from the funnel (shown in the figure) by blowing into the narrow end.[14]

Let's assume our student thinks that one can indeed blow the filter paper out of the funnel. What is he thinking?

He might be thinking about the details of the picture, wondering about the gap between the filter and the funnel. Or is that the thickness of the glass funnel? He may reflect back on the last time he made coffee, trying to remember if the filter came to a point as shown. Why won't the filter rest on the bottom of the funnel? Maybe it will slide out. Should he worry about the friction between the filter and the funnel at the line of contact? He worries about picking up the funnel and holding it horizontal so that the filter does not fall out without blowing at all. Maybe you're supposed to hold it vertical? Could the filter paper be wet, sticking to the funnel, preventing its removal by blowing? Does the weight and thickness of the filter paper matter? Its porosity?

Say now we give our student a clue: We suggest he go to his textbook and read up on Bernoulli's equation. He does so and comes to recognize that when he blows down the tube that, although the air-stream will exert a pressure on the filter's surface between the filter and the inside surface of the funnel, the resultant of this pressure –which would tend to push the filter to the right, out of the funnel – may be less than the pressure exerted by the atmosphere on the inside surface of the filter, the resultant of which would tend to push the filter back into the funnel. So maybe you can't blow the filter out of the funnel after all.

At this point, we might say he has identified his misconception. But has he? What was his misconception? That the filter was wet? That the funnel was made of glass? That it was held vertically? That the porosity of the paper might matter? It would be more accurate, more meaningful to say he had no conception of the phenomenon at all rather than that he had misconceived its functioning. It's the fog of the unknown that he is struggling with, a rambling array of possible interpretations of the text, including the figure, leading to nowhere in particular, not a misconception. What's wrong here is the reification of that which doesn't exist. It is all of a piece with the knowledge-as-stuff, the mind as a bucket image metaphor. In a less strident tone, we might only advance a more literal rendering of his failing, e.g., that he *miss*ed the proper *conception,* allowing that he had no concept. But how far did he miss it by? This again, is a queer way of speaking.

It is possible to identify a misconception if we work at it with him in a Socratic exchange, leading him on to construct a faulty, common-sense explanation of the phenomenon. But in this case, the misconception is as much our making as it is his. Misconceptions are a dialogic construction, a social production. Discovering a misconception is like constructing failure only in this case the system gone haywire is the student and the process requires a dialogue between the teacher and the errant student. Some speak of bringing the misconception to the surface so it can be exposed and disposed of properly, to be replaced with the correct conception, but this is, at least in this scenario, not the case. While this is a valid educational strategy; it can be very effective in bringing the student to see the world of fluid flow as the hydrodynamicist sees that world, it is a mistake to claim that the student was possessed with a misconception.

In our attempts to listen carefully and analyze the stories students tell, we need to understand that the picture of the funnel and filter is not a picture of a real funnel and filter but of one within a possible world of funnels and filters and streamlines and pressures and forces and continuum and time and motion. It is an ideal type within this world; a filter of zero, but impenetrable, thickness, of frictionless surfaces and steady blowing. It looks like something real but it is not. Our student's failing is an inability to imagine, to think in this object world. Rather than speak of a misconception its better to claim that he has not learned to see in this way; he has not learned the language of hyrdrodynamics.

That language is about steady, incompressible fluid flow, of gauge- and absolute pressures, of "head" and streamlines and mass flow. This is a world, a "game" which particular persons inhabit and play. The game has rules, a proper language - all that which constitutes a way of seeing, of thinking, and fixing legitimate moves. When we ask our student to explain why he can't blow the filter out of the funnel we are asking him to join in and play the game without telling him the rules. The language of hydrodynamics looks like English but it is not common sense English. (Language here is more than words, it is culture, social norms, ways of seeing the world). It's no wonder he doesn't think right. But once he claims to know the language, then if he errs it is legitimate to use the label - "misconception". In this case, it might take very little prompting for the student to recognize he was mistaken.

As in attempting to make use of history when our predecessors seemed to err, so too in attempting to understand the foreign ways of thought and methods of our students is problematic. The problem is in large part one of language; the students must recognize they are learning a new one; we, at the same time, in our traditional ways of developing theory, trying to make it accessible to our students, must be sensitive to the disjunctions in meaning cited above.

Normally in our elaboration of hydrodynamic theory we derive a sequence of mathematical relationships, proceeding from certain fundamental axioms, absent any narrative about funnels, or filters altogether. Our statements would be descriptive, not imperative. We would sketch a picture of a flowing fluid in a quite general and arbitrary form, showing streamlines, but it would be minimalist, consisting of a control volume much like the picture we have seen of a rigid body showing five particles. We would introduce the pressure at the head of the flow and downstream, velocity of flow at any point along a streamline. We would stay as close as possible to the analytical confines of the assumptions (incompressible, steady), concepts (pressure, velocity) and principles (momentum, conservation of mass) in unfolding the science.

The problem is, that if this is to be more than an exercise in applied mathematics, we must make reference to the furniture of the big world of fluids and connect it up with the formal analysis. In so doing, in talking about pipes and funnels, water jets and reservoirs, it is impossible to remain uncontaminated by common sense, by ordinary allusions expressed often, more problematically, in metaphorical language. This is a necessity; the mathematical relationships to the neophyte are just mathematical relationships which they might be very well prepared to manipulate and re-arrange in accord with the world of mathematics but this does not take us to where we want to go. We need the narrative of fluids, filters, and atmospheric pressure to get there. The challenge is to bring the student to see as we see, a special world, described in ordinary English, but distant in terms of "objective" meaning. We ought not expect them to make the translation", the reading we do effortlessly within the special world of Bernoulli. The problem of language is real.

Spending time with students in this way presents other challenges. Let's presume that our student doesn't stop with recognition of the applicability of Bernoulli's theory. Sensing he is on the right track, he seeks to apply Bernoulli's equation which relates the pressure to the velocity of flow along a stream line in an airstream. He knows he has to show that the pressure exerted by the stream of air flowing past the cone, between the cone and the filter outer surface, is everywhere less than the pressure on the other side of the filter due to the atmosphere.

But now he wonders: If there is a force to the left, why doesn't the filter move to close the gap until there is no gap, forcing the filter up firmly against the funnel? Then in this case, there is no flow. But how can you *not have flow* and yet *have flow*? Maybe the porosity of the filter paper does matter.

He muses further; perhaps one has to blow hard enough initially to move the filter out enough so there is some flow so that the atmospheric pressure will push it back. Or maybe this is why they show the gap between the filter and the funnel. When you blow, the filter moves back against the funnel; then it moves out, to the right again, so maybe the filter oscillates back and forth, fluttering in the airstream.

Now he really is in trouble; he is speaking the right language but now he is thinking too deeply; he has gone too far, introducing concepts which are beyond the scope of the course. He has a conception of how things go and why one might not be able to blow the filter out of the funnel. In fact, he might even have tried blowing a filter out of the funnel and discovered that indeed, he hears a rattling sound which he claims supports his conjecture. Unfortunately, the analysis of the fluttering behavior of a filter in a funnel is an advanced topic, one that requires more than hydrodynamic theory. Bernoulli's equation does not contain sufficient information to describe this dynamic, perhaps aeroelastic, phenomenon.

Here, strangely enough, it *is* legitimate to say he has erred, and this in two ways: In the first place, while he knows the rules of the game, he has not conceived the abstract representation as it was meant to be seen. His image is too sophisticated. Second, broadening the context of the exercise to include appropriate strategies the student might take to succeed in the course, he has misjudged what is required to earn the approval of faculty. Either way, matters in this case have moved out of control: Faculty may wonder how they are going to make up for the time lost in addressing the interesting provocations of certain students, yet sense that to cut students off would let an opportunity for learning slide away. One simply has to draw the line somewhere, acknowledge the legitimacy of the student's thinking, questioning, and move on.

Using open-ended exercises.

While the study of history and student's thinking in the raw can be splendid vehicles for engaging our new fundamentals, they both fail to attend to a very important feature of engineering thinking and doing; they fail to address the context of contemporary practice. History has context, and that certainly must be respected, but it is not contemporary. Dealing with misconceptions is dealing with ideas, but ideas of a sole individual alone, irrespective of any particular context. Introducing open-ended exercises can, on the other hand, bring context to the fore.

If our new enriched set of fundamentals are taken seriously, it's clear that there has to be significant change in what goes on in the classroom both in our engagement of students in the material of our course and throughout the whole curriculum. This is not a question of changing the content to "cover" this new material or adding more courses. It is a question about questions, about what are legitimate questions, about attitudes toward content, about boundaries on our object-world subject matter.

I have already described how this might be accomplished. I have already described innovations faculty have pursued in problem-based learning, in hands on experiences, in the use of open-ended exercises. These can be the vehicles for the infusion we seek. Take, for example, the use of open-ended, design-type exercises in an engineering science course.

A truly open-ended task invites the student to deal with ambiguity and uncertainty; to be aware of how different contexts may demand different strategies even though the technical "problem" may appear the same. Students learn how to judge what resources to bring to bear in different settings – know when a crude model and estimate will suffice, when a detailed and sophisticated computer analysis is required. Degree and form of justification becomes a matter of context.

In an open-ended situation, if done right, students must appropriate the problem on their own terms. No longer is the quest to produce the answer the faculty has in his head or found in the back of the textbook. This is not to say that "anything goes": While the context allows that there may be more than a unique solution and a single approach, some solutions will be better than others, some worse, some outright wrong. Faculty provide guidance, leading students through what may often appear as a very muddled situation. It is their responsibility to judge the amount of effort the task will require and set expectations accordingly. They set constraints, but not over constrain; provoke thought but not dictate a method; respond to student questions but not with absolutist answers. Open-ended exercises, if done right, serve as a powerful vehicle for active learning, for engaging the student in a way not possible sticking to the traditional prescription of single answer problems. If done right they allow the student to deal with anomalies, to address incongruities, to conceive of all sorts of possibilities, to improvise the unheard-of.

They learn also about an important ingredient of engineering thought not touched upon by Scheffler, namely how to negotiate their differences when their proposals differ from and/or conflict with those of a partner working on a common project. They begin to see that designing is a social process.

Only by means of this sort might we open up the context of the student experiences to allow and encourage critical discussion and reflection on the problem at hand. This discussion can include normative issues as well as instrumental questions about technical concepts, principles and methods. In this way we can make the shift from teaching to learning, from talking to listening, from derivation to critical discussion and dialogue and do so throughout the hard core of the engineering/science curriculum.

As an example of the transformation that is possible to effect, I present two exercises, both of which I have used in an undergraduate, introductory course in engineering mechanics. One is a traditional, single-answer problem. The other is an open-ended design task.

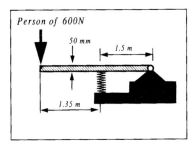

Figure 5.1 A Traditional Exercise

The traditional exercise is meant to teach the student the importance of the principles of static equilibrium, compatibility of deformation and the role of material properties in the analysis of the deflections of, and internal forces within, an elastic system. These are the traditional fundamentals.

The problem is stated thus:

> A wood diving board is hinged at one end and supported 1.5m from this
> end by a spring with a constant of 35 kN/m. How much will the spring
> deflect if a young man weighing 600 N stands at the end of the board?

There is but a single answer which, if the student is on top of the material of the course, should take no more than a half hour at most to produce. This presumes he is competent in the language of mechanics, i.e., recognizes the wiggly line as a linear elastic spring, understands that the circle at the right hand of the board is a frictionless pin, knows enough to assume the board itself is rigid (to analyze the situation if the board is not rigid comes later in the course) and is thoroughly at home with the metric system.

The second exercise presents very much the same picture but otherwise is dramatically different. The assignment is as follows:

> You are responsible for the design of a complete line of diving boards
> within a firm that markets and sells worldwide. Sketch a rudimentary
> design of a generic board. List performance criteria your product must
> satisfy. Include on your list those features which determine the perfor-
> mance of the board. Focusing on the dynamic response of the system,
> explore how those features might be sized to give your design the right
> feel...

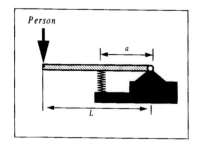

Figure 5.2 An Open-Ended Exercise

Now the dimensions of the board are not provided but are to be determined, or "sized". The written "problem" statement does not, can not, stand alone. It is essential that students and faculty discuss (negotiate) the problem requirements and expectations. Now the student must join in the formulation of the exercise, think about "performance criteria" such as the weight of the person using the product. Or is it persons? If so, what range of weights ought to be accommodated?

The fundamental principles of mechanics are still very much at the center of the task but the context is much richer. A discussion of "dynamic response" is necessary. Here I have the students imagine they are at the end of a diving board, testing its "bounce". They estimate a frequency of bounce. We talk about "equivalent stiffness" - the relationship between the deflection at the tip of the board and the weight of the person, about whether one can neglect the weight of the board or its flexibility. What about safety?

The students keep a journal. Since there is no single, unique response, the methods the student apply, the assumptions they make and their strategy for comparing different possible designs need to be evaluated. In this, faculty expectations must accord with a reasonable estimate of time available for the students to "complete" the exercise. I normally require that their journals be handed in within a week's time.

A warning: An open-ended exercise can be closed down if one puts the tinted glasses back on, if one wavers in the light of the student's aggressive questioning - their taking control - or in the weight of the task of reading through their reports. (The task admits of more than a single answer so evaluation must turn to focus on process and methods). This must be resisted.

This single example, taken from but a single course, appears not to say much about educational reform if our gaze remains fixed on content, what goes in the syllabus, or what courses should constitute a curriculum. But if we look from another perspective we might see that it addresses the fundamentals of engineering and engineering education across the board, through and through, irrespective of discipline and level - indeed, irrespective of content. That is the point. One can imagine how the traditional content in any subject might be recast in another mode, one that encourages the sort of open ended enquiry and sensitivity to context advocated here.

When we stop and think about it, those characteristics and aspects of learning which Scheffler claims are necessary to problem-solving are just the ones we value as researchers and in our engineering practice in the rough. We find, thus, we are not in another world but a familiar one indeed. But it is one we have to see properly, from a perspective that rejects the instrumental ontology, one that sees the fullness of all that constitutes engineering thought and practice.

Notes.

1. Masi, B. *The Impact of Strategic Initiatives on Structure, Governance and Resource Allocation in U.S Research Universities*, Ph.D. Thesis, MIT, 2001.

2. Bucciarelli, L.L., "Educating the Learning Practitioner", Invited Lecturer, SEFI Annual Conference, Vienna, Sept. 1996.

3. Note: this is an "ideal type"; an abstract construction I have inferred from faculty writings and our pronouncements about curriculum.

4. This is not to claim that the 50's types did not do risky things, start their own companies, go to law school or medical school or even become marketing gurus. I am talking about governing myths, values and beliefs, not attempting to depict reality otherwise.

5. Felder, et al (2000), in "The Future of Engineering Education II Teaching Methods That Work" *Chem. Engr. Education*, **34** (1), pp. 26-39.

6. ibid., p. 33.

7. ibid., p. 27

8. *Soft* characterizes a trait or property - or an idea, an argument, a proposition - that can not be measured by instrumental or quantitative methods. Engineers deal with the *hard*.

9. Felder, op.cit, p. 30.

10. Scheffler, I., *In Praise Of The Cognitive Emotions And Other Essays In The Philosophy Of Education*, New York: Routledge, 1991, p. 72

11. ibid., p. 73

12. ibid., p. 93

13. Brush, S., "Should the Teaching of the History of Science be rated X?", *Science*, 183, 1974, pp. 1164-72.

14. Halliday, D., and Resnick, R., *Fundamentals of Physics*, 2nd. ed., John Wiley & Sons, NY. 1986. p.313

6

Extrapolation

While engineering and philosophy may be worlds apart, I hope the reader at this point would agree that the questions philosophers address and the analyses they pursue are relevant to engineering - that the ways in which engineers think through their designs, deal with the malfunctioning of their productions, and teach the young are better understood if we bring a philosophical perspective to bear. Ontological questions about the existential qualities of the representations engineers make and use in their work are important and need critique and evaluation if our designing, our diagnosing, our teaching is to improve. Otherwise we risk presuming too much of our productions, conferring on them a solidity and robustness unwarranted in fact which eventually will lead to error. Epistemological questions about the status of knowledge claims and the form and tenor of their justification likewise are important to address for, without this sort of enquiry, we remained boxed in, constrained and limited to see engineering solely as reductive technique, materialist representation and manipulation.

At the same time, in exploring this terrain we are led away from philosophy; we discover that what engineers do, the reasons they do what they do, and their justification for believing the latter are good and true - all of this is historically and culturally contingent, not fixed by the dictates of science and empirical fact alone. Out of this comes a picture of both the possibilities and limitations of engineering thinking, what it can accomplish, what it cannot do - a picture with more depth and color than that we draw when constrained to paint only in instrumental tones.

Key to my analysis is the notion of "object-worlds" - the idea that different participants in design see the object of design differently depending upon their competencies, responsibilities and their technical interests. The ontological status of the object then becomes problematic. It can mean, can be, different things to different people. The same object, say a prismatic bar, to the structural engineer is a cantilever beam while to the person responsible for ensuring that the system does not overheat, it is a radiating appendage. This multifaceted character of the object of design - one object, multiple object-worlds - ensures that instrumental rationality does not suffice in design, nor in diagnostics and ought not be the sole thrust of engineering education.

In the course of my essay, I have noted the ways we divide up the world when we come to speak of technique. Within the design process we split up the design task and establish interfaces, hoping to keep control of the project; linguistic and disciplinary boundaries divide object-worlds; we posit gaps between structure and function, between the hard and the soft, between the sure and the uncertain, between product and social context. Maintaining these disjunctions depends upon seeing the artifacts of engineers - both ideas and objects - as hard, deterministic, permanent, totally rule bound, and material. I have argued that this view is deficient. These divisions are as much an artifact of culture as they are a consequence of science and the material richness of our world.

Take for example, the status of engineering theories, so critical to the justification of designs, the explanation of failure as well as so central in engineering education. These, expressed in mathematical form as in a textbook, appear formal, wholly analytic, universal in their applicability, quantitative and as sure as the numbers they produce - seemingly all on their own. They provide the structure - what Pirsig called the underlying form - which ensures the proper functioning of the products of engineering design. But their expression and manipulation-in-use presents a more confused picture: They are more supple, more plastic than they first appear. In designing and in diagnostic activities they are reconstructed, pared and sculpted to the particular problem at hand. The legitimacy of such moves depends upon what is acknowledged, perhaps only implicitly, as accepted, conventional practice, which, in turn, is a social affair.

The distinction between *structure* and *function* of engineering productions is likewise not as sharp as some would believe. In engineering, design, diagnostics and in teaching - every statement of structure, of the properties of an object other than those we can call familiar or vulgar, requires an extended story constructed in accord with the prevailing perspective and the particular needs of the task at hand[1]. This usually entails speaking about some aspect of the "performance" or "behavior" of the object.

A beam—for example, is more than a prismatic bar exhibiting certain dimensions and made of a certain material. If these were the only properties the structural engineer had available to work with, we would have neither bridges nor buildings. For this we need at least a *modulus of elasticity* and a number which defines when the material will fail. These additional, specialized, object-world properties, have meaning only when related to the performance of the object, i.e., its function as a beam. The modulus of elasticity requires talking about its linear, elastic *behavior* - how it will deflect an amount proportional to the amount of weight at the end; how it will return to its original configuration if the weight is removed.

To define when the beam will break, we refer to another feature of performance. In Galileo's story, he relates the weight at failure to that weight which when suspended from the same beam hung vertically would cause failure. Today, we define a property - the "yield stress" - whose numerical value we can look up in a table. But, like Galileo's measure, this value only has meaning as a consequence of the engineering profession's setting out of a standard test requiring special machines, special instrumentation, and special conventions over the years. Object world properties are socially contingent thus in two ways; in the ways they depend upon the particular narratives engineers construct in their day to day work and in the conventions which define methods to ascertain the values of properties.

In the chapter on malfunction and diagnosing failure, I tested the notion that engineers can claim certainty with respect to their designs. I concluded no: all engineering products and systems remain under-determined. The boundary between the sure and the unsure is never so clear that we can predict with certainty how our productions will perform. We never can anticipate all contexts of use.

In the same vein, products, like information, can be read differently by different persons and be put to use in different ways. An engineered artifact can function in a variety of ways, some unthought-of by participants in the design process themselves. The variety of function is inherent in the indeterminacy and the underdetermined nature of the design process. This, in turn, derives in large part from the challenge of bringing the different visions of the design professed by different participants into a coherent harmony.

That a design remains under-determined also means that any claim of technical perfection is unjustified. Within an object world, one may indeed be able to find an optimum design with respect to some subset of properties and with respect to the object's performance appropriate within that world. We can claim that algorithmic perfection is possible within an object world. In the design of the whole however, now taking our context as that of the project or the firm, this is not possible. There is no instrumental synthesis which can dictate perfection without negotiation of participants proposals, claims and demands. Preferences, technical preferences, are negotiated over object worlds. In the big world, however, attributing value or quality to a technical product is always a social process - reflected in my analysis of the social construction of failure. The boundary between a "good" and a "bad" production becomes diffuse.

Accepting this more complex vision of engineering thought and practice suggests an extrapolation is in order - one to tease out its implications for the way we think about the role of technology in our lives. First off, it strikes me as unhealthy to continue to refer to the "impacts" of technology on society as if technology was one *thing* - hard, structuring, determined, sure - and society *another* - soft, functioning, indeterminate, uncertain. Likewise to think that technology "has a life of its own" may be in order within an object-world of replicating automaton, but it is romantic nonsense to think and talk this way out here in the big world. So too to imagine we can perfect a missile defense shield, that we can profit from the genetic manipulation of life at all levels without occasioning significant collateral damage, or that we can convince every scientist that global warming is upon us before it is too late to do anything about it - all of this wishful thinking. It follows from a seriously flawed vision of technology, one that sets it apart and aloof, distant and seemingly out of reach of ordinary people. As citizens, we ought to know and do better.

fin

Notes

1. Bucciarelli, L.L., "Object and Social Artifact in Engineering Design", *The Empirical Turn in the Philosophy of Technology*, Kroes, P. & Mejiers, A., (eds). Elsiver Science, 2000.

Index

A

ABET 81, 86
Active learning 82, 83, 84, 94
Analogy 18
Analysis
    instrumental 6, 84
Analytic
    language as 18
    within object world 31
Anomaly 65
Artifact 30, 70
Assessment
    instrumental 26

B

Beam
    properties 100
Behavior
    of object 31
    of product 27
    of structures 56, 59, 87, 100
    unethical 41
Behaviorism
    social agents 7
Belief 27
    and knowledge 25
    fundamental, in teaching 79
    grounds for 45
    in a fix 41
Big world 12
Black Box 12
Brainstorming 33

C

Cantilever beam 22
    as lever 57, 88
    failure 58, 88
    of Galileo 57
    stresses within 6
Cause
    and effect 23, 45, 88
    of error 36
    of failure 36, 38, 41
Ceteris paribus 7, 23, 52
Cognition
    malfunction 90
    situated 8

Computer

Computer
    "bug" 26
    in education 83
    information technology 81
    methods 13
    program 9, 13, 47
    program "de-bugging" 43
    reduction 18
    simulation 28
    software 26
    tools 5, 13
Concepts
    and laws 53
    and principles 46, 53, 72, 95
    of equilibrium 64
    ontological status 17
    theoretical 58
Conjecture
    historical 62
    phenomenological 78
Constraint
    equations of 15
Context
    of development 46
    of learning 87
    of student experience 95
    of use 36, 101
    social/historical 46, 55, 62, 70, 100
Control
    in experiment 67
Cosmology 6
Counter factual 20
Creativity 48
Culture
    and language 17

D

Design
    as instrumental process 13, 16
    as social process 9, 11, 13, 17, 23, 94
    context 19, 25, 31, 41
    function 6
    interface requirement 21, 30
    knowledge 27
    malfunction 41, 99
    negotiation 9, 20, 22
    object 11, 12, 15, 19
    parameters 12, 18, 19, 21, 30